BIOCHEMISTRY RESEARCH TRENDS

GLYCOGEN

STRUCTURE, FUNCTIONS IN THE BODY AND ROLE IN DISEASE

BIOCHEMISTRY RESEARCH TRENDS

Additional books in this series can be found on Nova's website under the Series tab.

Additional e-books in this series can be found on Nova's website under the e-book tab.

BIOCHEMISTRY RESEARCH TRENDS

GLYCOGEN

STRUCTURE, FUNCTIONS IN THE BODY AND ROLE IN DISEASE

PEDRO L. WEISS
AND
BRIAN D. FAULKNER
EDITORS

New York

For permission to use material from this book please contact us:
Telephone 631-231-7269; Fax 631-231-8175
Web Site: http://www.novapublishers.com

Additional color graphics may be available in the e-book version of this book.

Library of Congress Cataloging-in-Publication Data

ISBN: 978-1-62948-395-5

Library of Congress Control Number: 2013951587

Published by Nova Science Publishers, Inc. † New York

CONTENTS

PREFACE

Glycogen is a multibranched polysaccharide that serves as the cellular energy storage in animals, fungi, and humans. In this book, the authors present research in the study of the structure, functions in the body and role in disease of glycogen. Topics discussed include the epidemiology, pathophysiology and genetics of Glycogen Storage Disease Type II (GSDII); glycogen metabolism during chronic liver diseases; muscle and liver glycogen in helminth-infected fish; and glycogen metabolism enzymes as molecular targets for drug development.

Chapter 1 - Glycogen Storage Disease Type II (GSDII) is an autosomal recessive lysosomal storage disease due to alpha acid glucosidase (GAA) deficiency *(GAA* gene, 17q25.2-q25.3), a ubiquitary lysosomal enzyme that degrades glygogen into glucose. GSDII prevalence varies from 1:40000 to 1:146000. Most authors consider only the infantile and late onset form, the latter being a large continuum.

More than 350 pathogenic mutations have been found and some of them concern the majority of patients.

In the neonatal period GAA mobilizes glycogen from the liver and later it avoids its accumulation in the lysosomes but its exact role(s) in adulthood remains uncertain.

Several stages of glycogen accumulation have been described and autophagia probably plays a central role in the process leading to muscle tissue damage. Glycogen accumulates in numerous organs including skeletal muscle, heart, liver, smooth muscle and the anterior horn of the spinal cord.

In the infantile form, the clinical picture is a floppy infant with severe respiratory distress, heart and liver failure. Death commonly occurs before one year of life. In some cases the first symptoms can occur later but by definition, before two years of life. While the disease generally leads to severe disability and premature death due to respiratory failure, the late onset form represents a spectrum of phenotypes from the mildest to the most severe.

Compared to other genetically-determined myopathies, GSDII causes a disproportionate respiratory failure mainly due to diaphragm weakness. Other particular phenotypical features were described as tongue enlargement, ptosis, hypoacusia, arterial abnormalities, incontinence or hydromyelia.

No consensus is currently admitted concerning the possible complex relationship between genotype and phenotype.

The diagnosis, in late onset form, can be challenging as ancillary tests can be uninformative. It is based on a low level of GAA activity and can be confirmed by the demonstration of biallelic pathogenic mutation.

Muscle biopsy can show typical abnormalities such as vacuolar myopathy with PAS positive and/or highly reactive acid phosphatase vacuoles.

The management of patients includes physiotherapy, the detection and care of respiratory failure. Enzyme replacement therapy (ERT) has shown its efficiency concerning disability progression and life expectancy but the response to the treatment varies among the patients.

In conclusion, GSDII is a rare multisystemic lysosomal storage disease that displays a wide spectrum of phenotypes and prognoses. The recent progresses of diagnosis and follow-up skills, and, the new care approaches including ERT are promising to improve the prognosis of patients.

Chapter 2 - The chapter focuses on the metabolism of glycogen and glucose during chronic hepatitides and cirrhosis of various etiology. Chronic hepatitides and their final stage, liver cirrhosis, are common and severe human diseases. They cause the impairment of the liver's morphological structure as well as various changes in the metabolism of carbohydrates, including that of glycogen, which is reflected in the glucostatic function of the liver. In particular, carbohydrate metabolism during cirrhosis is similar to that during diabetes, whose main feature is tissue intolerance to glucose. The chapter mostly presents the results of the authors' original research. Carbohydrate metabolism in the normal and the pathologically changed liver is examined during starvation and at various stages of the trophic cycle. The chapter contains a detailed discussion of regulatory mechanisms of glycogen metabolism in the normal and the cirrhotic liver, which help maintain the glucose homeostasis during starvation, in the absorptive and post-absorptive periods of the trophic cycle. Data on the activity of several key enzymes of glycolysis, gluconeogenesis, glycogenesis and glycogenolysis are given. Much attention is given to the glucose and glycogen metabolism under conditions of changing severity of the liver's chronic impairment.

Chapter 3 - The influence of trematode and cestode infections on the glycogen content in the muscles and liver of various fish species was studied. The glycogen content was found to increase in the muscles of young cyprinid fishes (ide *Leuciscus idus*, dace *Leuciscus leuciscus*, goldfish *Carassius auratus*, and roach *Rutilus rutilus*) following the experimental infection with the metacercariae of the trematode *Apophallus muehlingi*. The same was true for the roach juveniles naturally infected with another trematode *Posthodiplostomum cuticola*. These changes could be explained by a decrease in swimming activity of fish due to parasite-induced muscle injuries. As an example, a significant drop in spontaneous motor activity was shown for the crucian carp *Carassius carassius* after 30-day maintenance with the trematode-shedding mollusks, which may be a possible reason for glycogen accumulation in the muscles of infected fish.

At the same time, glycogen concentrations in the muscles and liver declined when the fish (perch *Perca fluviatilis*, bream *Abramis brama*, and sabrefish *Pelecus cultratus*) were infected with the plerocercoids of some cestodes (*Triaenophorus nodulosus* and *Ligula intestinalis*). Reduced glycogen content in the liver testifies to a dysfunction of this organ in the infected fish and, as a consequence, leads to a decrease in the activity of major digestive enzymes (proteinases and glycosidases) in the intestine. Thus, the infection with trematodes and cestodes differently affects the glycogen content in the fish tissues and, respectively, might cause different consequences for animal health.

Chapter 4 - Glycogen, the cellular energy store, is associated with a set of enzymes controlling its metabolism in a complex and dynamic way. Although glycogen metabolism has been thorough investigated, the regulatory mechanisms governing this process are not fully understood to date. New ideas have emerged in an effort to elucidate the complications of regulation of glycogen metabolism, from altered phosphorylation status and enzyme translocation to the role of intracellular compartmentalization and heterogeneity of glycogen stores as a regulator of metabolic pathways, and signalling transduction and lysosomal disposal of glycogen. The correlation of glycogen metabolism to many metabolic disorders has targeted many of the enzymes involved, for the development of drugs for a number of metabolic diseases, predominantly diabetes mellitus but also cancer. The notion of rational drug design emerged as the most appropriate modern interdisciplinary approach. This approach involves the close collaboration of diverse fields from bioinformatics and organic synthesis to structural biology (X-ray crystallography, NMR and EM) and physiology. In this review the authors will attempt to highlight major aspects and recent progress in the research for new drugs targeting enzymes playing a dominant role in the degradation rather than the synthetic glycogen pathway, such as glycogen phopshorylase and glycogen phosphorylase kinase.

In: Glycogen
Editors: P.L. Weiss and B.D. Faulkner
ISBN: 978-1-62948-395-5

Chapter 1

Glycogen Storage Disease Type II: Expanding the Comprehension of Clinical and Genetic Spectrums of a Generalized Disease

Gauthier Remiche[1,2,*], Maurizio Moggio[3] and Giacomo P. Comi[4]

[1]Fonds Erasme pour la Recherche Médicale
Université Libre de Bruxelles (ULB), Bruxelles, Belgium
[2]Department of Neurology, Hôpital Erasme,
Université Libre de Bruxelles (ULB), Brussels, Belgium
[3]Neuromuscular and Rare Diseases Unit, IRCCS Foundation Ca' Granda
Ospedale Maggiore Policlinico, Milan, Dino Ferrari Centre,
University of Milan, Italy
[4]Dino Ferrari Centre, Neuroscience Section, Department of
Pathophysiology and Transplantation (DEPT), University of Milan,
Neurology Unit, IRCCS Foundation Ca' Granda Ospedale Maggiore
Policlinico, Milan, Italy

* Corresponding author: Gauthier Remiche, MD. Neurologie, ULB Erasme, 808 Route de Lennik, 1070 Bruxelles, Belgium. Phone: 32 2 555 39 92. Fax: 32 2 555 39 42. E-mail: gauthier.remiche@erasme.ulb.ac.be.

ABSTRACT

Glycogen Storage Disease Type II (GSDII) is an autosomal recessive lysosomal storage disease due to alpha acid glucosidase (GAA) deficiency *(GAA* gene, 17q25.2-q25.3), a ubiquitary lysosomal enzyme that degrades glygogen into glucose. GSDII prevalence varies from 1:40000 to 1:146000. Most authors consider only the infantile and late onset form, the latter being a large continuum.

More than 350 pathogenic mutations have been found and some of them concern the majority of patients.

In the neonatal period GAA mobilizes glycogen from the liver and later it avoids its accumulation in the lysosomes but its exact role(s) in adulthood remains uncertain.

Several stages of glycogen accumulation have been described and autophagia probably plays a central role in the process leading to muscle tissue damage. Glycogen accumulates in numerous organs including skeletal muscle, heart, liver, smooth muscle and the anterior horn of the spinal cord.

In the infantile form, the clinical picture is a floppy infant with severe respiratory distress, heart and liver failure. Death commonly occurs before one year of life. In some cases the first symptoms can occur later but by definition, before two years of life. While the disease generally leads to severe disability and premature death due to respiratory failure, the late onset form represents a spectrum of phenotypes from the mildest to the most severe.

Compared to other genetically-determined myopathies, GSDII causes a disproportionate respiratory failure mainly due to diaphragm weakness. Other particular phenotypical features were described as tongue enlargement, ptosis, hypoacusia, arterial abnormalities, incontinence or hydromyelia.

No consensus is currently admitted concerning the possible complex relationship between genotype and phenotype.

The diagnosis, in late onset form, can be challenging as ancillary tests can be uninformative. It is based on a low level of GAA activity and can be confirmed by the demonstration of biallelic pathogenic mutation.

Muscle biopsy can show typical abnormalities such as vacuolar myopathy with PAS positive and/or highly reactive acid phosphatase vacuoles.

The management of patients includes physiotherapy, the detection and care of respiratory failure. Enzyme replacement therapy (ERT) has shown its efficiency concerning disability progression and life expectancy but the response to the treatment varies among the patients.

In conclusion, GSDII is a rare multisystemic lysosomal storage disease that displays a wide spectrum of phenotypes and prognoses. The recent progresses of diagnosis and follow-up skills, and, the new care approaches including ERT are promising to improve the prognosis of patients.

1. HISTORY AND DEFINITIONS

Glycogen Storage Disease Type II was firstly described in 1932 by Joannes Cassianus Pompe [1]. He described a case of an infantile fatal cardiac hypertrophy. Autopsy, at seven months of age, disclosed vacuolar glycogen infiltration of the heart.

More than twenty years later, in 1955, Christian de Duve described for the first time the lysosomes [2] that was rewarded by the Nobel Prize in 1974. It was only in 1965 that Henri-Gery Hers discovered the deficient enzyme in GSDII from which emerged the concept of storage disease [3]. In 1986, Martiniuk et al. identified the gene encoding for the GAA enzyme [4]. Since then, publications provide more and more evidence that recombinant enzyme replacement therapy can provide an improvement in the evolution of disease [5-10].

Acid alpha-1,4-glucosidase (GAA) (synonym : acid maltase) is a lysosomal enzyme that degrades glycogen, maltose or oligosaccharides to glucose.

GSDII is an autosomal recessive generalized storage disease due to a biochemical defect in the lysosomal acid alpha-1,4-glucosidase that is encoded by the *GAA* gene [11-14]. It can cause a large panel of phenotype's severity and ages of onset.

Many organs can be involved such as heart failure, muscle weakness, respiratory failure, liver function impairment and many others that will be detailed in the clinical description section.

2. EPIDEMIOLOGY

Prevalence of GSDII varies between different populations ranging from 1:40 000 to 1:146 000 [15-17]. Its estimation varies regarding the geographic location: 1/57 000 in The Netherlands (adult form) [16], from 1:131 600 to 1:83 000 in the USA [15] or 1:146 000 in Australia [18]. The proportion of new GSDII patients born per year was estimated as high as 1/39 000 (including neonatal and late onset forms) [15].

3. Pathophysiology and Genetics

3.1. Acid Alpha-1,4-Glucosidase or Acid Maltase

Acid alpha-1,4-glucosidase (GAA) is a lysosomal enzyme. It is optimally active when the pH range is between 4 and 5 and degrades maltose, oligosaccharides or glycogen into glucose [14]. Hydrolyzing α-1,4 and α-1,6 glucidic linkages it is able to completely depolymerize the glycogen into glucose residues. This enzyme is found in many tissues including heart, liver, muscle, gut, kidney, and leukocytes [14].

It is produced as a high molecular weight precursor and will be maturated following several steps that will lead to the mature enzyme.

A signal peptide (N-terminal extremity) allows the entrance into the endoplasmic reticulum, after that, it is removed and the protein will be glycosylated with high-mannose residues before going to the Golgi apparatus. In the latter, it will be phosphorylated, glycosylated and will acquire a membrane marker recognizing the mannose-6-phosphate membrane receptor [14]. The enzyme discloses seven sites of glycosylation (two being eliminated during the maturation). GAA leaves the Golgi apparatus in vacuoles that fuse with the reserve of lysosome enzymes and form the primary lysosomes [14]. The enzyme will herein proceed to additional maturation steps, namely C- and N-terminal cleavages, which will provide the specificity of its action to glycogen [14].

The exact role of this enzyme has not yet been completely elucidated. As we observe a large accumulation of glycogen in GAA deficient cells, in cultured fibroblasts the contribution of this enzyme seems extremely low related to the role of the cytoplasmic neural maltases [14].

Acid alpha-1,4-glucosidase gene (*GAA*) (OMIM 606800) location is on locus q25.2-q25.3 of the chromosome 17. Its size is 3.6Kb and it is composed of 20 exons.

Most of the mutations are privates (point mutations, deletion, insertions, inversions, etc.), and more than 350 pathogenic mutations have been reported [19].

3.2. Neutral Maltases

Neutral Maltases are ubiquitous cytosolic enzymes with alpha-glucosidase activity (AB or II, C and maltase-glucoamylase). Type C releases glucose from cytosolic glycogen [14]. One of them, namely the AB or II subtype, encoded by the *GANB* gene on chromosome 11, is the most active enzyme when the artificial substrate 4-methylumbelliferyl-α-D-glucopyranoside (4MUG) is used on tissue to assess alpha-glucosidase activity [14]. This enzyme, if not inhibited by acarbose, is responsible for inappropriate interpretation of GAA residual activity when assayed on leukocytes [14].

Several questions arise about the pathophysiology of GSDII, among the most important ones, we can cite the comprehension of the mechanism leading to cell destruction and the question related to origin of the wide clinical, biochemical and genetic heterogeneity.

3.3. Mechanisms Leading to Cell Degeneration

Contrary to "hepatic" glycogen storage diseases (GSD) such as type 0, I, III, IV, VI, IX or XI, in GSDII, the hepatic mobilization of glycogen is not compromised so that patients are not exposed to the risk of hypoglycemia [14, 20].

If it is well known that the primary defect causing the clinical picture of GSDII is the excessive intralysosomal glycogen storage [12, 14], little is currently known regarding the event's cascade that leads to muscle dysfunction and muscle weakness.

Damage to muscle fibers is due to the accumulation of large amounts of glycogen that cause displacement, replacement and compression of other normal organelles of the cell [14]. As glycogen is a hydrophilic molecule it takes a large proportion of the available volume of the cytoplasm. This reduction of available space in the cell should be the *primum movens* leading to muscle cell damage in GSDII. Secondary to this, compensatory mechanisms are activated as a partial degradation of glycogen in lysosomes, exocytosis and the elimination of free glycogen by phosphorylations. However, these mechanisms are not sufficient to overcome the increasing glycogen accumulation and the progression of the disease.

Excessive lysosomal activity is suggested by the hypertrophy of the lysosomal system that is also seen in unaffected muscle fibers and by the presence of large autophagic vacuoles in muscle fibers severely affected by the disease [14].

These vacuoles resemble the autophagic vacuoles that are seen in chloroquine-induced myopathies. In this case, autophagy is triggered by the lysosomotropism of the molecule that will secondarily lead to the destruction of this organelle [14, 21]. GAA knock out (KO) mice and GSD II patients disclose a global vacuolar dysfunction including trafficking pathways, endocytosis and autophagy [22, 23]. In GSDII, secondary to GAA deficiency, autophagy will become inappropriate and uncontrolled. It will participate in muscle cell degeneration. Moreover, it is also suggested that lysosomal enzymes leak or are released from secondary lysosomes due to their rupture.

The presence of lysosomal enzymes in the cytosol will have dramatic consequences such as the destruction of other organelles and the inactivation of cytosolic enzymes that includes enzymes implied in the cytosolic glycogenolysis [14]. The latter mechanism will consequently participate in the process of glycogen accumulation. Ultrastructural observations are in line with this theory as gaps are seen in lysosomal glycogen sacs. However we can oppose this theory in that the lysosomal enzyme should be less active in the cytosol because they are optimally active at low pH such as in the lysosomes [14]. The group of Raben et al and other groups have demonstrated the potential central role of autophagic vacuoles on the GDSII pathophysiologic process from the enzyme defect to the muscle weakness [24, 25, 22, 23].

Following works of Winkel et al [26], the degree of muscle impairment and the degree of autophagy could be an important issue used to predict the response to enzyme replacement therapy (ERT).

3.4. Possible Explanations/Hypotheses for the Clinical Variability

Clinical and biochemical differences observed between patients are due to the clinical and genetic heterogeneity of the disease. Several mechanisms could explain these differences:

- *The hypothesis of the residual enzyme activity*: residual enzyme activity should be linked to the severity of tissue impairment. Indeed, in the infantile form, no residual activity is detectable while in the late onset form it is only reduced. This theory is only partially valid for several reasons. The first is that the residual activity is determined by an artificial substrate (it does not reflect exactly what happens in physiologic conditions); the second is that patients with the same residual activity can have a highly variable phenotype and the third is that we can see patients harboring an infantile form with a detectable persistent residual activity and late onset patients without detectable activity. Moreover, no explanations are currently provided to explain why the level of excessive glycogen storage in infantile or adult onset patients can vary highly following the sampled muscle or the part of the sampled muscle [14].

- *The molecular differences related to the type of deficiency [27,14]:*
  - The enzyme is produced but its catalytic activity is severely reduced or absent.
  - The enzyme is not produced.
  - The quantity is reduced and its activity is too low.
  - The hypothesis of the existence of an isozyme for GAA.
- *The neutral maltase hypothesis:* in the infantile form of the disease, skeletal muscle tissue analysis showed that neutral maltase is deficient but such impairment was not observed in adult patients. Then, we can suppose that the latter enzyme could partially compensate for the GAA deficiency. Only one of the neutral maltase is able to hydrolyze glycogen, namely type C, but some parts of the population are asymptomatic carriers of a null allele for this enzyme. That would also be an explanation for the clinical variability that can be observed between patients disclosing the same genotypes. The presence of the null allele could enhance the cytosolic glycogen clearance deficiency and result in an increase of autophagia.
- *The deficiency of synthesis, phosphorylation and maturation of GAA:* Several methods are used to assess the biochemical defect on fibroblasts. The incorporation of L-[4,5^{3}H] leucine or [^{35}S] methionine can provide an index related to the rate of production of the enzyme. The use of [2-^{3}H] mannose provides the possibility to obtain a measurement of its glycosylation and the incorporation of free [^{32}P] allows the measurement the mannose's phosphorylation. These techniques give the possibility to define five different biochemical defects related to several clinical forms of the disease:
  - The 110 kDa precursor is not synthesized.
  - The 110 kDa precursor is synthesized but its amount is reduced while the maturation is normal.
  - The 110 kDa precursor is normally synthesized but degraded before its transfer for maturation into the Gogi appartatus.
  - Synthesis and glycosylation are normal but the phosphorylation and subsequent intracellular processes are deficient.

- All the processes of synthesis are normally performed but the final cleavage by proteolysis that leads to the mature form of the enzyme (76 kDa) is blocked.

Then, there is no correlation between biochemical and clinical phenotypes. These works unfortunately were not yet correlated with molecular genetics analyses [14].

3.5. Genomic Structure and cDNA of the *GAA* Gene

The size of cDNA is 3.6 kb including 2856 bp of coding sequence. Three regions can be distinguished: a 105.37 kDa coding for 952 unglycosylated amino acids, one untranslated on 3' extremity of a size of 550 bp before the terminating codon and another untranslated region of 440 bp. The total size of the gene is 25 kb and it contains 20 exons. The first is a 5'untranslated sequence that is separated from the following exon by a 2.7 kb intron containing the ATG initiation codon. It also contains regulation sequences that could play a role in regulation of the transcription. The second and the last exon are relatively large (578 and 607 bp), others having a size comprised between 85 and 197 bp. The second exon codes for the signal peptide, the prosequence for the precursor and the first 61 AA of the 76 kDa enzyme precursor. The last exon codes only 60 bp. The promoter is located just before the first exon and a sequence of 400 bp is sufficient to promote the expression. To reach the maturity stage, the enzyme must follow post-translational modifications. A prosequence is cleaved in the endoplasmic reticulum; the enzyme will be glycosylated afterwards and will be phosphorylated on mannoses residues in order to confer the targeting to lysosomes. In the lysosomes it will be cleaved to acquire its mature form of 76 and 70 kDa [14]. Several mutations are private but some molecular defects occur in most of the patients and may segregate with their geographic origin [28, 17, 12].

4. Clinical Description

4.1. Clinical Forms

Historically, phenotypes were classified following the age of onset of first symptoms, the age of death, the velocity of evolution and the affected organs.

Two different classifications exist, one in two groups and the other separating three subclasses. Supporters of the three groups' classification advance two mains arguments. Firstly, some patients with infancy onset can survive to the second decade of life and, secondly, patients who present first symptoms in the first years of life and die during the second decade are different than patients who present mild symptoms during the seventh decade [14]. Another distinction is to only separate infantile and late onset forms, which will be detailed later. The *infantile form* begins in the first days of life. Main symptoms progress rapidly and include hypotonia, heart, liver and tongue enlargement. Respiratory failure and feeding difficulties are frequent and death occurs generally before the age of 2, due to cardiorespiratory failure. Glycogen accumulation concerns the main part of tissue including the central and peripheral nervous system. While this entity is clinically homogenous, it is associated with various biochemical and genetic findings. The *childhood onset or juvenile form* begins during childhood with a proximal myopathy. Patients have delayed motor milestones, a preferential failure of respiratory muscle can occur as well as calves hypertrophy (leading to differential diagnosis for muscle dystrophy). The evolution is slow and death is due to the respiratory failure. Few patients survive over the second decade. Cardiomegaly, hepatomegaly and macroglossia are infrequent. Rupture of basilar trunk aneurisms were described and autopsy disclosed abnormal storage material in the lysosomes of vascular walls myocytes. The *adult onset form* begins after the age of 20. It leads to progressive axial and proximal weakness mimicking a muscle dystrophy or a myositis. Because a clear-cut separation cannot definitely be made between the juvenile-onset form of the disease and the adult-onset form of the disease, despite the recent proposition of Gungor et al. [29], the main current trend in the literature is to consider only the infantile/classical and late onset form [11, 30, 31, 12, 17, 32-36]. In the late onset form a wide clinical heterogeneity has been observed [14, 17, 28, 33, 34, 37, 38] and divided the late onset form into several distinct clinical categories seems to us difficult to apply but this point is currently being debated.

4.2. Involved Organs and Clinical Consequences (Late Onset Form)

4.2.A. Striated Skeletal Muscle

A selective non-homogeneous pattern of muscle involvement is described in GSDII (e.g. predominance of atrophy on the posterior compartment of the femoral muscles) [39, 14]. The level of atrophy is generally correlated to the weakness [14]. Previously we systematically studied the pattern of muscle strength impairment in a cohort of 19 late-onset GSDII patients [40] (Table 1).

Table 1. Mean modified medical research council (mMRC) strength scores for muscle or muscle group in late onset glycogen storage type II patients

muscle or muscle group	mean mMRC score (%)
hip extension (P)	67.6%
iliopsoas (S)	70.3%
hip abductor (LD)	72.5%
thigh abductor (S)	78.9%
external rotator of the shoulder (P)	84.1%
thigh extensor (S)	84.2%
knee flexor (P)	84.7%
external rotation of the arm (S)	85.6%
pectoralis (S)	86.3%
internal rotation of the arm (S)	86.7%
thumb abductor (S)	86.8%
infraspinatus (S)	87.4%
triceps (S)	90.0%
biceps brachialis (S)	90.3%
thigh adductor (S)	90.5%
head flexor (dd)	92.5%
deltoid (S)	94.2%
quadriceps femoris (S)	95.0%
tibialis posterior (S)	96.1%
hand flexor (S)	97.4%
head extensor (P)	98.2%
wrist extensor (S)	98.4%
muscle or muscle group	mean mMRC score (%)
tibialis anterior (S)	98.7%
wrist flexor (S)	99.2%
food flexor (P)	99.4%
head extensor (S)	99.4%
elbow extensor (dd)	99.7%
head flexor (S)	100.0%

Mean of left and right modified medical research council (mMRC) strength scores for the different muscles or muscle groups assessed in a cohort of 19 patients with late onset glycogen storage type II [40]. Position during strength assessment: P: prone, S: seated, dd: dorsal decubitus and LD: lateral decubitus.

The groups are classified from the most compromised to the most preserved.

It showed mainly an involvement of the pelvic and the scapular girdles, a moderate weakness of the muscle of arms and the thigh and a sparing of the distal muscles of the forearm and the leg, but a relative sparing of some neck muscle.

The remarkable muscle MRI imaging study of Carlier et al. [41] confirmed that some muscles were earlier and more severely affected such as the systematic axial muscle involvement particularly concerning neck and trunk extensors (lumbar extensors) and abdominal belt followed by scapular girdle (*subscapularis* and *anterior serratus*) and pelvic girdle (*adductor magnus*). Tongue muscles involvement was consistently observed and suggestive of GSDII.

To the opposite, other muscles were spared as most of facial muscles, all other neck muscles than the neck extensors and intercostal muscles. The muscles of the tight were heterogeneously affected without anterior or posterior predominant involvement, but *rectus femoris*, *gracilis* and *sartorius* were relatively spared.

Our clinical observations [40] and previous publications of our group [42] are in line with the main part of the observations made by Carlier et al. [41].

Moreover, the severe and early involvement of diaphragm muscle has been largely described [17, 43, 44].

4.2.B. Respiratory Striated Skeletal Muscle

Some cases have been described as having respiratory failure as the first symptom of the disease, [14] but its frequency is described from rare cases [45, 31, 30, 11, 46] to one third or more of non infantile-onset cases [14, 12, 47].

Because the severity of respiratory involvement is not always related to the degree of skeletal muscles weakness [15], respiratory failure can occur insidiously (or sometimes acutely) even in the early stages of the disease, so that patients may ultimately become dependent on mechanical ventilator support when limb muscles are not severely compromised [43, 45, 31, 48, 49, 34].

Symptoms can be related to sleep disordered breathing with nocturnal hypercapnic hypoventilation such as nocturnal or morning headache, day sleepiness or nausea. Due to restrictive syndrome and diaphragm failure orthopnea and exertional dyspnea can be also seen [50, 51, 14] .

As a result, respiratory failure, mainly due to the disproportionate involvement of the diaphragm, is the first cause of death in late onset GSDII [17, 43, 14].

Therefore, correct monitoring of respiratory function in late onset GSDII patients is an important prognostic factor for the timely initiation, not only of mechanical ventilation, but also of enzyme replacement therapy [43]. The assessment of respiratory function by spirometry discloses a restrictive syndrome that in early stages can only be seen in a supine position. Nocturnal oximetry and analyses of blood gases will show a hypoxemia-hypercapnia [14].

Respiratory muscle outcomes in late onset GSDII have been noninvasively investigated through the measurements of maximal inspiratory and expiratory pressures, which deteriorate over time, indicating that both inspiratory and expiratory muscles are affected [52-54].

Diaphragmatic weakness is defined as a fall of forced vital capacity (FVC) greater than 25% between seated and supine position [55].

A new promising technique, namely, opto-electronic plethysmography has been proved to allow a reliable non-invasive selective assessment of respiratory muscle function with specific evaluation of the diaphragm and was demonstrated as useful to discriminate diaphragm failure in GSDII patients [44]. It could be a future alternative or additional tool for the routine respiratory assessment and follow-up of GSDII and other neuromuscular patients.

4.2.C. Multisystem Involvement

As GSDII is a multisystem disorder, numerous publications have underlined that numerous organs can be involved by glycogen accumulation and with variable clinical consequences. GSDII murine models [56] and human studies [57, 58] have shown that glycogen accumulation reaches almost all tissues. Table 2 summarizes the main tissues that can be involved in GSDII and their clinical consequences. Generally neither significant *cardiomegaly* nor severe *hepatomegaly* is found in the adult onset form of the disease [14].

While cardiac failure is not a part of the clinical picture in late onset GSDII, mild ventricle hypertrophy, reduced contractility or Wolff-Parkinson-White syndrome can be encountered [59]. Liver abnormalities (such as hepatomegaly or steatosis) were reported in various proportions [60, 34].

Aneurism and vascular arterial malformations such as dolichoectasia of the basilar Artery (Figure 1), aorta dilatation, ectasia of internal carotids have been described in GSDII for several years [61, 62, 34, 51, 59, 14]. Aneurisms can sometimes rupture [14] and other vascular malformations have sometimes led to transient ischemic attack or cranial nerve palsy [61].

Table 2. The main involved tissues in glycogen storage type II are listed here in front of the corresponding clinical consequences

Involved Tissues	Clinical Consequences	Clinical form		References	
		infantile	late onset	infantile	late onset
girdles and limbs striated muscles					
	proximal weakness	+++	++	14,67,69	14,77
	scapular winging	?	++	not found	14,59
axial striated muscles					
tongue	tongue enlargement	+	+	14	70,14,59,41
spine	scoliosis, hyperlordosis	+/-*	++	67	14,59
pelvic floor	incontinence	+	++	75	42
diaphragm	respiratory failure	++	++	14,67,69	17,43,14,44
facial striated muscles	facial weakness/ptosis	++	+	67-69	14,59
heart					
nodal tissue	Wolff-Parkinson-White syndrome	+	+	14,69	14,59
striated muscle	cardiac enlargement/cardiomyopathy and failure	+++	+/- **	14,57,67,69	77,14,59
smooth vascular muscle	dolicho basilar trunk, aneurisms + rupture	+	+	14,63,58	61,77,14,59
smooth digestive muscle	incontinence	+	++	75	42,77
anterior horn of the spine	unknown (neurogenic EMG, fibre type groupping on muscle biopsies)	+	+ (groups of atrophy)	14,57,58	14
periphereal nerves (Schwann cells)	unknown (neurogenic EMG, fibre type groupping on muscle biopsies)	+	+	14,58,75	77
central nervous system	unknown (infantile) ; hydromyelia (late onset)	?	+	not found	76,77
otoacoustic alterations, Corti's organ	deafness and otitis	+	+	71,72,69	73,74
liver	liver enlargement and failure for infantile onset	++	+ (enlargement only)	14	77
bulbar musculature and upper esophagus	dysphagia-dysphonia	++	+	119,68	77,120
consequences of muscle weakness	orthopedic	+/-***	++		31,34,59
due to heart failure, respiratory failure and bulbar weakness	failure to thrive	++	-	14,36	/

(*) rarely observed in few survival patients ERT treated; (**) left ventricle hypertrophy, no cardiac failure; (***) osteopenia

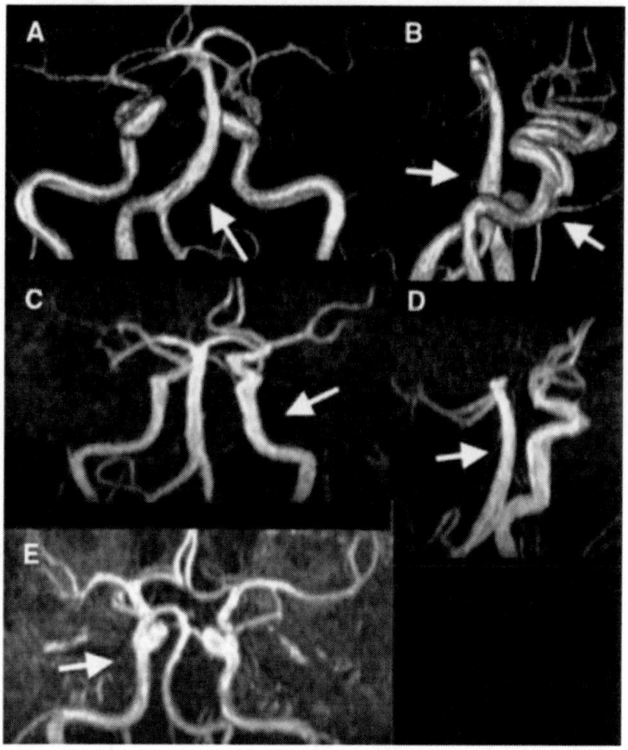

Figure 1. Magnetic resonance angiography late onset GSDII patients. Basilar artery dolichoectasia (a, arrow) and bilateral intracavernous carotid dolichoectasia (b, arrow) by 3D-time of flight (TOF). Basilar artery dolichoectasia and dilated intracavernous left carotid on (c, d arrows) by 3D-TOF sequence with maximum intensity projection (MIP) reconstruction: Dilated right carotid (e arrow) on 3D-TOF sequence with MIP reconstruction [61].

One case of aneurism in an infantile onset GSDII was described [63].

Some studies reported *facial weakness* and *ptosis* with variable frequencies in adults [64-66] and in infants [67-69].

Tongue enlargement is described in infantile onset GSDII [14] as well as in the late onset form. In the latter group, tongue weakness is a part of the clinical picture and is correlated to impaired muscle signals on MRI [70, 14, 59, 41].

Hearing alterations have been described in infantile onset as well as in late onset form of GSDII [71-74]. In the infantile form it concerns more than 90% of patients and it has a mixed origin: cochlear and conductive dysfunction [72, 69].

In the late onset form, the majority of patients deny subjective hearing disturbances but audiological evaluation reveals that 52.5% have hearing impairment (57% sensorineural, 33% conductive loss and 10% a mixed pattern) [73]. On the other hand, van der Beek et al. conducted a study on 58 adults with adult onset GSDII and they found that the prevalence of hearing impairment (21% of clinically relevant hearing loss) was not greater that what is seen in the general population. They concluded that hearing loss is not a specific feature of adult onset GSDII [74].

Incontinence was described as a debilitating underdiagnosed condition in late onset GSDII. It concerns all range of patient's age and affects urine, fecal or both types of continence. Its prevalence was assessed as high as 25% and some patients were improved by ERT [42]. This condition was also recently described in several cases of infantile onset GSDII survival thanks to ERT [75].

Orthopedic complications also occur frequently in the late onset form [31, 34, 59] and are scoliosis (77.3%), lumbar hyperlordosis (65.9%), rigid spine syndrome (15.9%), scapular winging (27.3%) (Figure 2), joint contractures (6.8%) and pes equines 2.3%) [59].

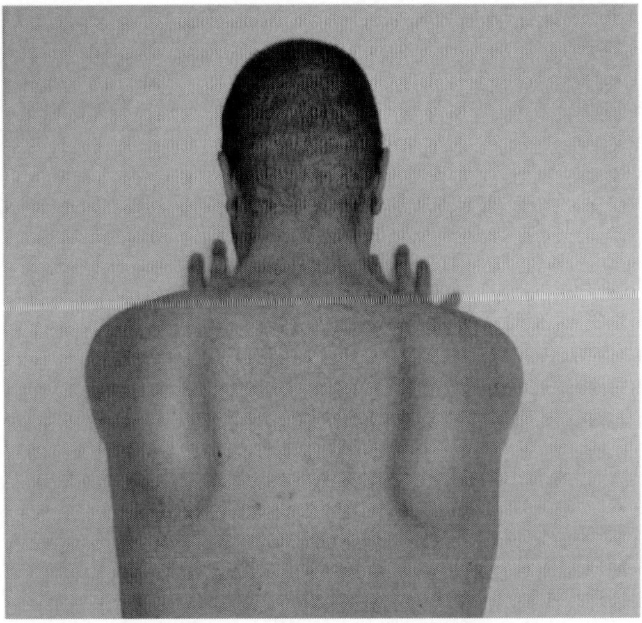

Figure 2. Scapular winging in a 40-year-old late onset male patient.

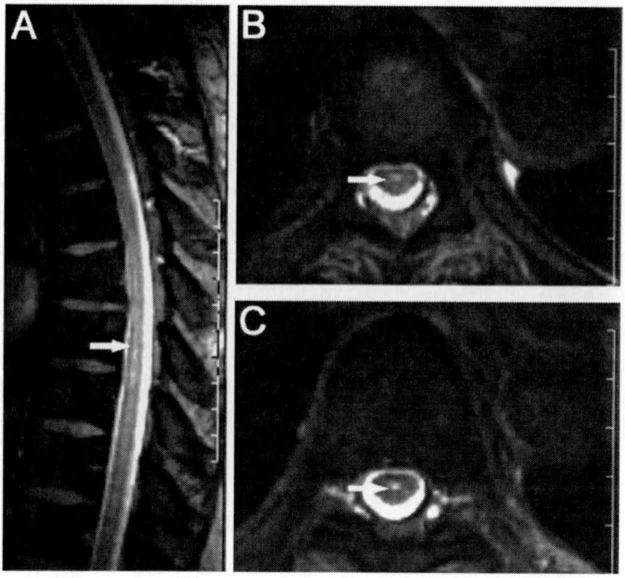

Figure 3. The spine magnetic resonance imaging (MRI) of a patient showing the dilatation of the central canal regarding vertebras Th6 to Th8. Because it is limited to central canal dilatation, it argues for the hydromyelia diagnosis. A: stagittal T2 MRI imaging; B and C: axial T2 MRI imaging. Arrows show the hydromyelic cavity [40].

Other rare organ involvement has been described recently as the presence or *hydromyelic cavity* in two patients without signs related to central nervous system impairment (Figure 3) [76, 77].

5. Ancillary Studies and Diagnosis

5.1. Blood Analyses

Blood analyses reveal a high creatine kinase (CK) level in most of the cases. Moreover aspartate aminotransferase and lactate dehydrogenase are increased in the different forms of GSDII [14]. CK level is generally more increased in infants but generally less than tenfold the upper limit of normal [14]. However, in the late onset form high levels can be found up to 15 fold [78] with a varying mean following the studies from 2.8 to 3.8 [65, 59, 78].

Normal CK levels do not exclude GSDII and can be found in the late onset form with variable percentages (from 9.1% to 29.2%) [79, 65, 78, 14].

On the other hand, high CK levels should lead one to consider late onset GSDII even in cases of asymptomatic or paucisymptomatic subjects as previously described [52, 65, 80, 30, 39, 81, 82]. Currently the reliable, rapid, non-invasive, cheap and easy to perform *dried blood spot (DBS) test* is available to assess GAA residual activity on leukocytes [83-88].

It has been used in order to screen GSDII in neonate patients in Taiwan [89] and its reliability has also been demonstrated in late onset patients [86, 82]. Currently the use of this test as a first-line test to diagnose GSDII is admitted and included in several recommendations [87, 85, 90, 88]. Unfortunately precise data about sensitivity and specificity related to well-defined populations are currently lacking. We can consider that this test, when found positive, can support the clinical diagnosis presumption. However, a definite confirmation needs the use of a gold standard method such as the GAA residual activity measurement on cultured fibroblasts [86, 33, 84].

5.2. Electromyography (EMG)

It reveals a myopathic pattern in the different forms of the disease, abnormalities being more salient in children. The presence of abnormalities can vary according to sampled muscle. Hyperexitability can be found such as myotonic discharges (pseudo-myotonia) that are more widespread in children than in adults [14]. They can be preferentially found in paraspinal and abdominal muscles (sometimes in adults, they are only detectable in paraspinal muscles) [14, 91]. Fibrillations and positive sharp waves as complex repetitive discharges can also be detected. These abnormalities of excessive electrical irritability are supportive to the diagnosis. The sensitivity of EMG is not absolute as 20% of EMG can be non-myopathic [65] (26.7% in children and 16.7% in adults) [91]. Neurogenic or mixed patterns can also be found; it could be due to the possible involvement of the anterior horn of the spine by glycogen storage.

5.3. Muscle Imaging

Muscle magnetic resonance imaging (MRI) can provide suggestive patterns of muscle involvement for late onset GSDII that could be helpful to evoke its diagnosis [41]. Suggestive patterns have been described in the clinical description section of this chapter.

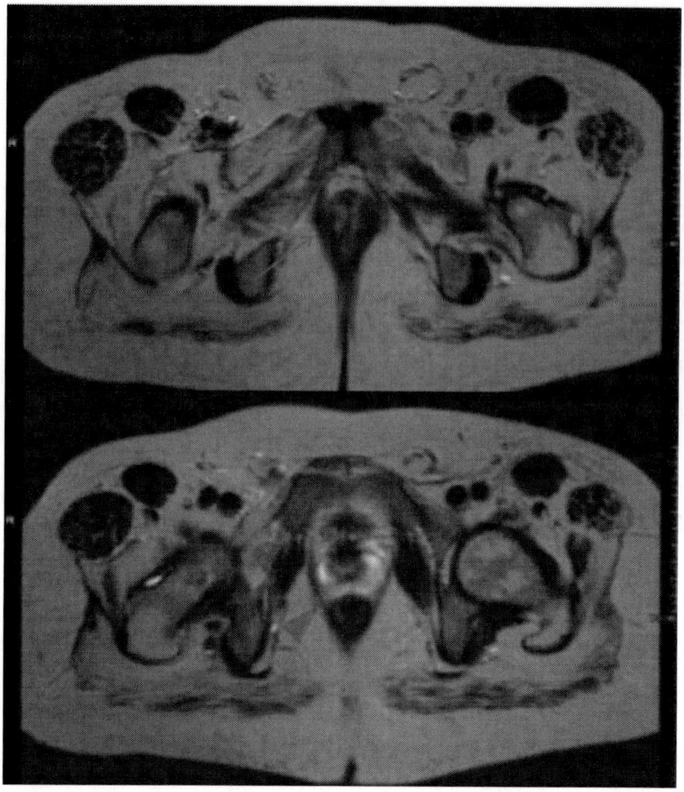

Figure 4. Pelvic muscle magnetic resonance imaging of a late onset GSDII patient showing the diffuse major fatty infiltration of pelvic floor muscles. Arrow shows the normal trophism of sphincter *ani externus* and the arrowhead shows the atrophy of *levator ani* muscle. Other muscles have severe fat replacement such as *gluteus maximus*.

Interestingly, pelvic floor muscle involvement, which is sometimes difficult to assess clinically, can be precisely assessed by MRI [42] (Figure 4).

5.4. Histopathology

Common findings in *light microscopy* of muscle biopsy are those found in vacuolar myopathies. In young children almost all fibers contain large vacuoles. As the age of onset increase, the vacuolation trend is less obvious and some fibers are spared. Generally, clinically relatively spared muscles have less vacuolated fibers. These vacuoles have high glycogen content and a strong positive reaction to acid phosphatase stain, which indicates their lysosomal origin (Figure 5). Periodic Acid and Shiff (PAS) method also shows a marked positive reaction especially in neonates (Figure 6) [14, 92].

The presence of rare vacuolated fibers associated to multiple acid phosphatase fibers must lead to suspect GSDII. The vacuolation can concern type I or type II fibers without preferential involvement. Necrosis and regeneration fibers can be found as well as splitting fibers and mononuclear endomysial or perivascular infiltration. In adult muscle a connective tissue replacement can be observed. Fiber atrophy and neuromuscular spindle alteration can also be found [14].

Denervation or neurogenic findings, such as fiber type grouping (Figure 7) or large groups of atrophic fibers, have been described in some patients [60, 14].

In the late onset form of the disease, muscle biopsy can be normal (25% in adult onset cases) [93] or aspecific [39, 80, 33, 60]. In the study of Laforêt et al. [39] 30% of the biopsies had neither vacuoles nor PAS positive material.

Recent recommendations of the American Academy of Electrodiagnostic Medicine (AAEM) conclude that a normal or aspecific muscle biopsy cannot exclude the diagnosis of GSDII (especially if performed in a slight clinically affected muscle) [87].

In *Electron microscopy*, glycogen accumulation is seen in four types of compartments.

"Lakes" of free glycogen can be disseminated in cytosol. One other compartment contains glycogen in a surrounded continuous or disrupted membrane structure (Figure 8) [21, 14, 92]. Sometimes, when muscle fibers are less compromised, these abnormalities can only been seen in electron microscopy (EM). In these spaces the size of the glycogen granules can be the same or smaller than those seen in the cytosol. That could be due to a partial degradation of them into the disrupted-membrane shaped spaces. We can hypothesize that this space are disrupted lysosomes [21]. Same spaces can be seen in satellites cells [14].

A third space containing glycogen is autophagia vacuoles and discloses a heterogeneous material [21]. A simple or a double membrane surrounds it. The fourth compartment corresponds to a transition between the two latter spaces described.

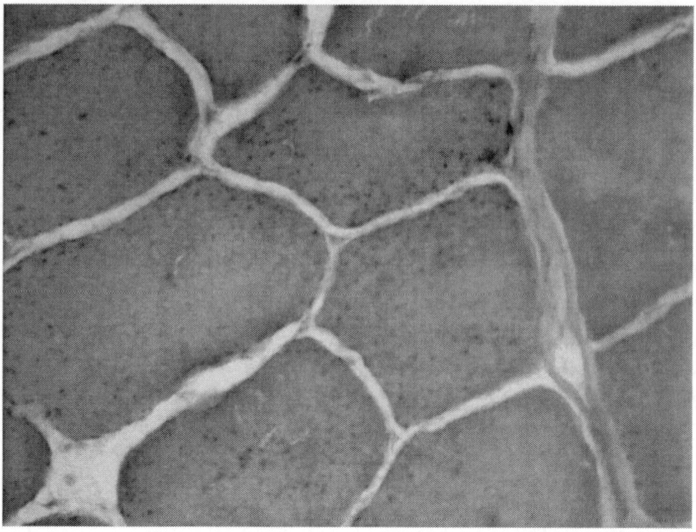

Figure 5. Light microscopy of muscle biopsy from a late onset GSDII patient. Acid phosphatase stain showing numerous muscle fibers harboring vacuols with lysosomal activation (red stained).

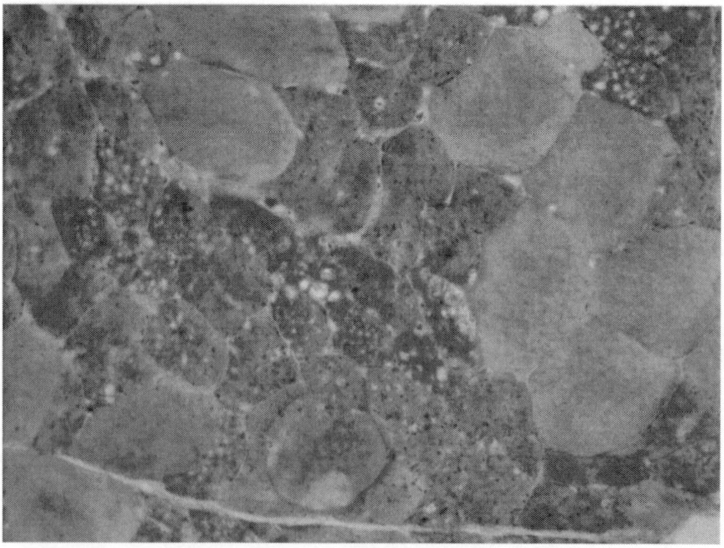

Figure 6. Light microscopy of muscle biopsy from a GSDII patient. Periodic Acid and Shiff stain showing an important glycogen accumulation among almost all muscle fibers.

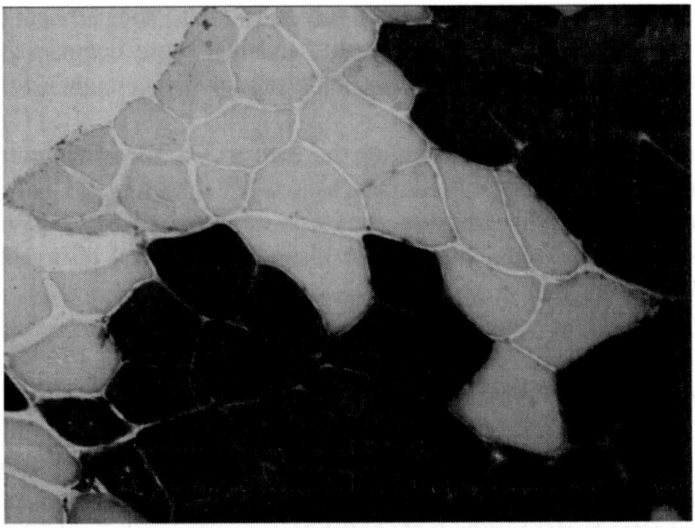

Figure 7. Light microscopy of muscle biopsy from a late onset GSDII patient. ATPase stain, at pH: 9.4. It shows a trend to fiber type grouping such as seen in neurogenic disorders including second neurone motor disease. White fibers: type I and dark fibers: type II.

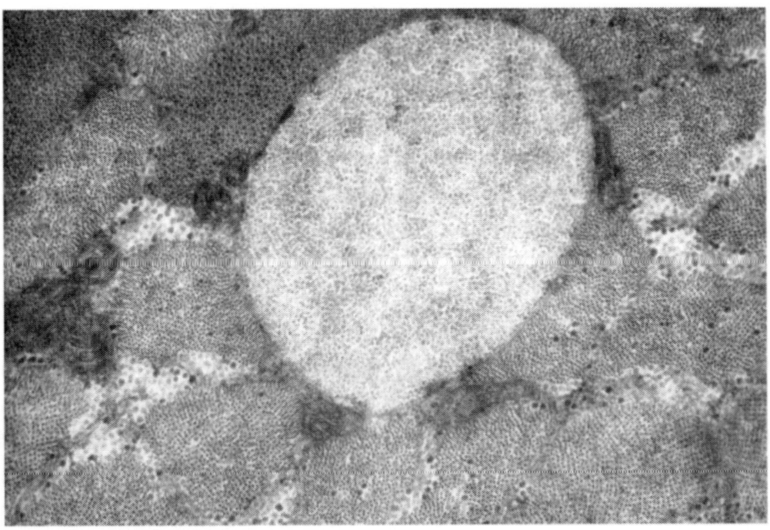

Figure 8. Electron microscopy of muscle biopsy from a late onset GSDII patient showing a lysosome filled with glycogen.

Moreover, muscle biopsy findings are not totally specific and should lead to a relatively small differential diagnosis.

Myopathies associated to autophagia cause a high acid phosphatase stain. It can be seen in toxic myopathies due to (hydroxy)chloroquine or colchicine therapy and in Danon disease or other rare X-linked myopathy with excessive autophagy [14, 99, 92].

7. The Management and Follow-up of Patients

GSDII is a multisystem disease and is best managed with a *multidisciplinary approach*. The team should be experienced in the management of such disease and led by neuromuscular or metabolic experienced and specialized physicians.

7.1. Cardiac Involvement

Cardiac risks are related to arrhythmia, cardiomyopathy and heart failure. Significant cardiac risks mainly concern the infantile form but severe arrhythmias such as Wolff-Parkinson-White have also been described in late onset forms [14, 59].

The infancy cardiac hypertrophy begins *in utero* and reaches a significant level between 4 and 8 weeks of life. A partial response to enzyme replacement therapy is observed. Conduction abnormalities including short P-R intervals could lead to tachyarrhythmia and sudden death. Therefore baseline and regular interval controls of 24-hour Holter-ECG monitoring and echocardiography are recommended. Infants disclosing cardiomyopathy should be specifically followed and treated by a specialized pediatric cardiologist [85].

7.2. Respiratory Management

GSDII leads to reduced *lung* volumes, weakening of cough and sleep-related breathing impairment. Respiratory involvement is not always related to the degree of skeletal muscle weakness [45] and respiratory failure can occur insidiously even in the early stages of the disease, so that patients may ultimately become dependent on mechanical ventilation support when limb muscles are not severely compromised [48].

As a result, respiratory failure, mainly due to the disproportionate involvement of the diaphragm, is the first cause of death in late onset GSDII [17, 43].

Therefore, correct monitoring of respiratory function in GSDII patients is an important prognostic factor [43].

Respiratory status assessment should be performed regularly. It will include a clinical evaluation of cough, the presence of wheezing or dyspnea, exercise tolerance, and for infants, and the ability to feed [85]. Spirometric measurements are only applicable in most children older than 2 but it will need encouragement and suitable measurement conditions between 2 and 6 years [100]. Pulmonary function tests should include measurements of VC, forced expiratory volume in 1 second (FEV1), maximal inspiratory pressure (MIP) and maximal expiratory pressure (MEP) [85]. Diaphragmatic weakness is defined as a fall of forced vital capacity (FVC) greater than 25% between the seated and supine position [55].

Opto-electronic plethysmography is a promising future routine test that provides a reliable non-invasive selective assessment of respiratory muscle function with specific evaluation of the diaphragm.

This method does not require advanced collaboration and is easily applicable to children [101, 102].

Gas exchanges should be assessed to detect progressive respiratory failure. Oximetry should be routinely performed. If normal, arterial blood gas analysis should not be systematically performed. Chest radiography will be helpful to detect atelectasis [85].

Assisted cough will improve the impaired secretion clearance due to muscle weakness and is recommended in order to preserve pulmonary function. It can be associated to the use of bronchodilators. Inspiratory muscle training may be used and pulmonary infections should be early and strongly treated. When hypoxia is present oxygenotherapy is appropriate.

Assisted ventilation is required for nocturnal hypoxia (nasal C-PAP in case of obstructive sleep apnea or Bi-level nocturnal noninvasive positive pressure ventilation for nocturnal hypoventilation). If hypoventilation is also present during daytime, continuous noninvasive positive pressure ventilation or tracheostomy with mechanical ventilation should be considered [85, 103].

7.3. Swallowing, Nutrition and Digestive Management

Due to several factors such as facial hypotonia, macroglossia, tongue and pharynx weakness, swallowing is limited in infantile-onset GSDII patients. In late onset patients, fatigue during chewing can occur that could lead to insufficient intake of nutrients.

Management of these patients (especially infantile onset or severe patient) by a nutrition team is then recommended [85, 103]. If a swallow study is abnormal, an adapted feeding regimen is needed and if a significant risk of aspiration is proven, nutrition should be switched to gavage feeding. For the infantile form of the disease or early childhood onset GSDII patients, growth parameters namely height, weight and head circumference must be regularly followed [85].

For late onset patients, a high protein diet associated with low carbohydrate nutrition have been proposed in order to tempt a reduction of the increasing glycogen accumulation and amino acid utilization. It was proposed in combination with aerobic exertion with encouraging results [104].

7.4. Physical Therapy

Physical therapy is an important part of the care of infantile as well as adult patients.

The aims are to preserve motor and physiological function and prevent secondary complications such as joint contractures. Strengthening should be performed with caution; if excessive, it is believed to be detrimental as is thought for most of the progressive myopathies.

Exercises should be gentle, aerobic and submaximal. Excessive resistive or eccentric exercise should be avoided [85].

7.5. Miscellaneous

The research of *incontinence* should be done for each patient because it is a debilitating condition that patients are reluctant to complain. Moreover, if present, specific investigation and treatment are needed and can potentially lead to significant improvement [42].

Contractures prevention can be performed by daily stretching and position correction by patients and family education as well as the use of *ad hoc* devices.

Osteopenia is frequently seen in GSDII and it should be regularly screened by dual energy x-ray absorptiometry (DEXA) and cured when needed [85].

For all patients, appropriate functionality and maximal level of *autonomy* should be an important goal. The use of mobility devices should be proposed when appropriate in order to maintain a maximal autonomy.

Infantile onset patients should be screened for *hearing loss*.

Immunization including seasonal influenza and pneumococcal *vaccination* and palivizumab administration in children during respiratory syncytial virus (RSV) season should be proposed.

Anesthesia should be avoided when possible and an experienced anesthetizing team is required to manage GSDII patients especially those with the infantile onset form.

A coordinated multidisciplinary specialized and experienced team should care for these patients and the assistance of a patient's association can be helpful for patients and their families.

As this disease (infantile or late onset form) implies severe and progressive disability, *psychosocial aspects* should always be investigated and the support of a social worker and/or psychologist should be available and proposed.

Genetic counseling must be offered to all patients or patient's family. The risk of recurrence when a child is affected is 25%. DNA analysis is needed to identify potential additional family members affected or carriers. *Prenatal diagnostic* testing is possible and several methods are available. Because it can be done by GAA activity measurement, the identification of mutations in the family is not necessarily required but is generally preferred. The combination of both approaches increases the reliability [85].

8. The Genotype-Phenotype Correlation Controversy

Some studies concluded to partial genotype-phenotype correlation [105,94,30,80,106] but numerous others [17, 32, 107, 38, 81, 35, 39, 37,11] did not. This apparent splitting could have several origins.

Firstly, some authors consider genotype-phenotype correlation because infantile onset patients harbor more frequent severe mutations on both alleles (such as c.2481+102_2646+31del or c.525delT).

Arguments against the existence of genotype-phenotype correlation could come from the fact that patients of a same family (siblings) can have very different phenotypes although they harbor the same mutation on both *GAA* alleles. Numerous reasons can explain it and were detailed in the section "*Pathophysiology and Genetics*". We can give an example as a reminder of the need for numerous other genes to produce a functional GAA enzyme (post-translational modifications such as glycosylations and phosphorylations) or the difference between individuals that could concern the process of autophagy [14].

Moreover, genomic characterization of patients do not assess the fact that some mutations can lead to variables consequences such as an inter-individual variability of the degree of retention of exon 2 due to the IVS1-32-13T>G or a variability of the level transcripts degradation produced from non-sense mRNA (as the c.525delT or other frameshift mutations)[108].

When a phenotypical variability is observed in non-related patients harboring the same mutation, single nucleotide polymorphism (SNPs) interindividual differences could modulate the effect of mutations on *GAA* gene expression leading to a difference in the level of severity clinically observed [14].

Some general trends concerning genotype-phenotype correlation have been considered: non-sense mutations, small insertions, deletion leading to frameshift and large deletion are severely deleterious. Missense mutations and those involving a splicing site have largely variable consequences: from null effect to partially deleterious with the persistence of a detectable GAA residual activity level [14].

In summary we can say that, generally, the association of two null mutations leads to an infantile phenotype, the association of two intermediate mutations (less severe) or the association between a less severe mutation and a null mutation will lead to a late onset phenotype [14].

Among the Caucasian population that presents an adult onset phenotype, the most frequently found mutation is IVS1-13T>G. This mutation is located on a splicing site and allows 5% of normal splicing. At least 70% of patients with this form of the disease harbor this mutation on one of the two alleles, the other mutation being a null mutation. Some rare patients homozygous for IVS1-13T>G have been described [109, 39, 110, 81, 111-113].

Most of the studies concerning genotype-phenotype correlation were performed on small samples, it makes it difficult to provide definite conclusions and these results are, by definition, exposed to type II error (lack of power).

Moreover, unfortunately the homogeneity of clinical assessment is lacking in the literature and this fact may impair our chance to compare different cohorts of patients and to detect potential subtle correlations involving clinical, biochemical and genetic aspects. The use of an international registry could help to overcome this limitation.

9. Experimental Models and Research

Animal models that are chemically induced or occurring naturally have been used for research. Acarbose is a competing enzyme for GAA and its repeated intraperitoneal injection in rats causes a generalized glycogen accumulation, but in this model, the cardiac or muscle storage remains slight. GSDII can occur naturally in several animal species such as Lapland dogs, Japanese quails and shorthorn or Brahma cattle from Australia [14].

Different murine models have been developed by molecular engineering that delete all or part of the *GAA* gene leading to *knockout* (KO) mice. These models correspond to a mixed of the infantile and the adult phenotype because muscle involvement looks like human late onset form of the disease; but, similar to infantile forms, KO mice harbor cardiac involvement. Interestingly the severity of the phenotype changes according to the genetic background of the mice. The latter observation could probably be explained by the presence of modifying genes [14].

These murine models have been used mainly to assess the effect of various therapeutic maneuvers such as enzyme replacement therapy (ERT), gene therapy and DNA injection.

We must remember that KO mice are not always a good reflection of what happens in human conditions [14].

GAA KO mice have also been largely used in order to improve our comprehension of GSDII pathophysiology. It was particularly used by the group of Raben et al. to obtain a better comprehension of the exact sequence of events and precise mechanisms leading to muscle degeneration that include autophagy for a large part [24, 22, 25, 23].

Numerous clinical researches are currently pending and concern various topics about pathophysiology, the variability of clinical picture and the benefit of therapies.

The Pompe Registry® collects clinical and genetic data about a great (n = 742) [114] number of GSDII patients but systematic phenotypical characterization is currently available only for a small subset of subjects.

10. THERAPIES

10.1. Enzyme Replacement Therapy (ERT)

ERT improves survival and clinical course in infantile-onset GSDII form (the *strictu sensu* Pompe disease), and has also more recently allowed new hope for late onset patients [12, 115, 116, 17, 79, 10].

However, it is extremely expensive and requires lifelong biweekly intravenous infusions.

The interindividual variability response to ERT is very important. At this time the comprehension of the factors that could predict or discriminate which patient will have the best response to ERT is not obvious. Numerous factors should be involved, and probably combined: age of onset, the degree of muscle impairment (e.g. fibrous-adipose replaced muscle is not supposed to be improved by ERT), genetic variability concerning GAA locus or non GAA locus genetic variability (modifying genes), environment interactions, presence of anti-recombinant GAA antibodies, etc. [14, 85].

No standardized validated criteria have been defined to select patients susceptible to have the major benefit for this treatment. Treatment of patients without significant symptoms (e.g. diagnosis made on basis of casual high CK level discovery secondary to screening when siblings are affected) or with a very late onset is still under discussion but some authors recommend not treating these subjects but to apply a close follow-up and initiate treatment only if symptoms or objective signs appear [103]. For patients who require the use of wheelchair and/or invasive ventilation ERT should be used for one year and further treatment should afterward be decided on a case-by-case basis [103].

10.2. Gene Therapy

Gene therapy has only provided clinical encouraging results in a very limited amount of diseases and patients [117, 118].

In a recessive disease due to the lack of enzyme functioning we can expect, as heterozygous carriers are asymptomatic, that 50% of normal level activity is sufficient to restore the normal function. Several viral vectors have been tried such as adeno and adeno-associated virus (AAV).

The first had the disadvantage of not providing permanent replacement and the second, integring the genome on an unpredictable site, could lead to oncogenesis [14].

10.3. Global Care of Patient

Care of patient should not be limited to these novel disease-specific approaches and must always include the supportive multidisciplinary care [85, 103] as detailed in the section *"The management and the follow-up of patients"* of this chapter.

Conclusion

Glycogen storage disease type II is a multisystemic storage disease that mainly leads to progressive limb weakness and respiratrory failure in the late onset form of the disorder. For patients with infantile onset, it first leads to cardiac failure.

Prognosis can be improved by multidisciplinary care of the patient and by enzyme replacement therapy.

Further research is needed to better understand the processes that lead to muscle degeneration and why such a wide variability is observed concerning the response to ERT. Consequently we hope to improve the efficacy of currently available cures and/or to develop new curative approaches for GSDII.

REFERENCES

[1] Pompe JC (1932) Over idiopathische hypertrophie van het hart. *Ned. Tijdschr. Geneeskd.* 76:304.

[2] De Duve C, Pressman BC, Gianetto R, Wattiaux R, Appelmans F (1955) Tissue fractionation studies. 6. Intracellular distribution patterns of enzymes in rat-liver tissue. *The Biochemical Journal* 60 (4):604-617.

[3] Hers HG (1963) alpha-Glucosidase deficiency in generalized glycogenstorage disease (Pompe's disease). *The Biochemical Journal* 86:11-16.

[4] Martiniuk F, Mehler M, Pellicer A, Tzall S, La Badie G, Hobart C, Ellenbogen A, Hirschhorn R (1986) Isolation of a cDNA for human acid alpha-glucosidase and detection of genetic heterogeneity for mRNA in three alpha-glucosidase-deficient patients. *Proc. Natl. Acad. Sci. U S A* 83 (24):9641-9644.

[5] Van der Ploeg AT, Kroos MA, Willemsen R, Brons NH, Reuser AJ (1991) Intravenous administration of phosphorylated acid alpha-glucosidase leads to uptake of enzyme in heart and skeletal muscle of mice. *J. Clin. Invest.* 87 (2):513-518. doi:10.1172/JCI115025.

[6] Kikuchi T, Yang HW, Pennybacker M, Ichihara N, Mizutani M, Van Hove JL, Chen YT (1998) Clinical and metabolic correction of pompe disease by enzyme therapy in acid maltase-deficient quail. *J. Clin. Invest.* 101 (4):827-833. doi:10.1172/JCI1722.

[7] Van den Hout H, Reuser AJ, Vulto AG, Loonen MC, Cromme-Dijkhuis A, Van der Ploeg AT (2000) Recombinant human alpha-glucosidase from rabbit milk in Pompe patients. *Lancet* 356 (9227):397-398.

[8] Van den Hout JM, Kamphoven JH, Winkel LP, Arts WF, De Klerk JB, Loonen MC, Vulto AG, Cromme-Dijkhuis A, Weisglas-Kuperus N, Hop W, Van Hirtum H, Van Diggelen OP, Boer M, Kroos MA, Van Doorn PA, Van der Voort E, Sibbles B, Van Corven EJ, Brakenhoff JP, Van Hove J, Smeitink JA, de Jong G, Reuser AJ, Van der Ploeg AT (2004) Long-term intravenous treatment of Pompe disease with recombinant human alpha-glucosidase from milk. *Pediatrics* 113 (5):e448-457.

[9] van der Ploeg AT, Clemens PR, Corzo D, Escolar DM, Florence J, Groeneveld GJ, Herson S, Kishnani PS, Laforet P, Lake SL, Lange DJ, Leshner RT, Mayhew JE, Morgan C, Nozaki K, Park DJ, Pestronk A, Rosenbloom B, Skrinar A, van Capelle CI, van der Beek NA, Wasserstein M, Zivkovic SA (2010) A randomized study of

alglucosidase alfa in late-onset Pompe's disease. *The New England Journal of Medicine* 362 (15):1396-1406. doi:10.1056/NEJMoa0909859.

[10] van der Ploeg AT, Barohn R, Carlson L, Charrow J, Clemens PR, Hopkin RJ, Kishnani PS, Laforet P, Morgan C, Nations S, Pestronk A, Plotkin H, Rosenbloom BE, Sims KB, Tsao E (2012) Open-label extension study following the Late-Onset Treatment Study (LOTS) of alglucosidase alfa. *Molecular genetics and metabolism* 107 (3):456-461. doi:10.1016/j.ymgme.2012.09.015.

[11] Kroos MA, Pomponio RJ, Hagemans ML, Keulemans JL, Phipps M, DeRiso M, Palmer RE, Ausems MG, Van der Beek NA, Van Diggelen OP, Halley DJ, Van der Ploeg AT, Reuser AJ (2007) Broad spectrum of Pompe disease in patients with the same c.-32-13T->G haplotype. *Neurology* 68 (2):110-115. doi:68/2/110 [pii]. 10.1212/01.wnl. 0000252798.25690.76.

[12] Raben N, Plotz P, Byrne BJ (2002) Acid alpha-glucosidase deficiency (glycogenosis type II, Pompe disease). *Curr. Mol. Med.* 2 (2):145-166.

[13] Engel AG, Gomez MR, Seybold ME, Lambert EH (1973) The spectrum and diagnosis of acid maltase deficiency. *Neurology* 23 (1):95-106.

[14] Engel AGH, R.; Huie, M. L. (2004) Acid Maltase Deficiency. In: Engel A (ed) *Myology,* vol 2. McGraw-Hill, Rochester, pp 1559-1586.

[15] Martiniuk F, Chen A, Mack A, Arvanitopoulos E, Chen Y, Rom WN, Codd WJ, Hanna B, Alcabes P, Raben N, Plotz P (1998) Carrier frequency for glycogen storage disease type II in New York and estimates of affected individuals born with the disease. *Am. J. Med. Genet.* 79 (1):69-72. doi:10.1002/(SICI)1096-8628(19980827)79:1<69::AID-AJMG16>3.0.CO;2-K [pii].

[16] Ausems MG, Verbiest J, Hermans MP, Kroos MA, Beemer FA, Wokke JH, Sandkuijl LA, Reuser AJ, van der Ploeg AT (1999) Frequency of glycogen storage disease type II in The Netherlands: implications for diagnosis and genetic counselling. *Eur. J. Hum. Genet.* 7 (6):713-716. doi:10.1038/sj.ejhg.5200367.

[17] van der Ploeg AT, Reuser AJ (2008) Pompe's disease. *Lancet* 372 (9646):1342-1353. doi:S0140-6736(08)61555-X [pii]. 10.1016/S0140-6736(08)61555-X.

[18] Meikle PJ, Hopwood JJ, Clague AE, Carey WF (1999) Prevalence of lysosomal storage disorders. *Jama* 281 (3):249-254.

[19] http://www.erasmusmc.nl/klinische_genetica/research/lijnen/pompe_center/3909417/.

[20] Laforêt PD, A.; Weinstein, G. ; Smit, P. A. (2012) The Glycogen Storage Diseases and Related disorders. In: Saudubray JvdB, G; Walter, J. H. (ed) *Inborn Metabolic Diseases*. Springer, Berlin, pp 115-139.

[21] Lewandowska E, Wierzba-Bobrowicz T, Rola R, Modzelewska J, Stepien T, Lugowska A, Pasennik E, Ryglewicz D (2008) Pathology of skeletal muscle cells in adult-onset glycogenosis type II (Pompe disease): ultrastructural study. *Folia Neuropathol.* 46 (2):123-133.

[22] Fukuda T, Roberts A, Ahearn M, Zaal K, Ralston E, Plotz PH, Raben N (2006) Autophagy and lysosomes in Pompe disease. *Autophagy* 2 (4):318-320. doi:2984 [pii].

[23] Raben N, Takikita S, Pittis MG, Bembi B, Marie SK, Roberts A, Page L, Kishnani PS, Schoser BG, Chien YH, Ralston E, Nagaraju K, Plotz PH (2007) Deconstructing Pompe disease by analyzing single muscle fibers: to see a world in a grain of sand. *Autophagy* 3 (6):546-552. doi:4591 [pii].

[24] Fukuda T, Ewan L, Bauer M, Mattaliano RJ, Zaal K, Ralston E, Plotz PH, Raben N (2006) Dysfunction of endocytic and autophagic pathways in a lysosomal storage disease. *Ann. Neurol.* 59 (4):700-708. doi:10.1002/ana.20807.

[25] Fukuda T, Ahearn M, Roberts A, Mattaliano RJ, Zaal K, Ralston E, Plotz PH, Raben N (2006) Autophagy and mistargeting of therapeutic enzyme in skeletal muscle in Pompe disease. *Mol. Ther.* 14 (6):831-839. doi:S1525-0016(06)01638-8 [pii]. 10.1016/j.ymthe.2006.08.009.

[26] Winkel LP, Kamphoven JH, van den Hout HJ, Severijnen LA, van Doorn PA, Reuser AJ, van der Ploeg AT (2003) Morphological changes in muscle tissue of patients with infantile Pompe's disease receiving enzyme replacement therapy. *Muscle Nerve* 27 (6):743-751. doi:10.1002/mus.10381.

[27] Beratis NG, LaBadie GU, Hirschhorn K (1983) Genetic heterogeneity in acid alpha-glucosidase deficiency. *American Journal of Human Genetics* 35 (1):21-33.

[28] Kroos MA, Van der Kraan M, Van Diggelen OP, Kleijer WJ, Reuser AJ, Van den Boogaard MJ, Ausems MG, Ploos van Amstel HK, Poenaru L, Nicolino M, et al. (1995) Glycogen storage disease type II: frequency of three common mutant alleles and their associated clinical phenotypes studied in 121 patients. *J. Med. Genet.* 32 (10):836-837.

[29] Gungor D, Reuser AJ (2013) How to describe the clinical spectrum in Pompe disease? *American Journal of Medical Genetics Part A* 161A (2):399-400. doi:10.1002/ajmg.a.35662.

[30] Ausems MG, ten Berg K, Beemer FA, Wokke JH (2000) Phenotypic expression of late-onset glycogen storage disease type II: identification of asymptomatic adults through family studies and review of reported families. *Neuromuscul. Disord.* 10 (7):467-471. doi:S0960-8966(00)00123-1 [pii].

[31] Hagemans ML, Winkel LP, Van Doorn PA, Hop WJ, Loonen MC, Reuser AJ, Van der Ploeg AT (2005) Clinical manifestation and natural course of late-onset Pompe's disease in 54 Dutch patients. *Brain* 128 (Pt 3):671-677. doi:awh384 [pii]. 10.1093/brain/awh384.

[32] Joshi PR, Glaser D, Schmidt S, Vorgerd M, Winterholler M, Eger K, Zierz S, Deschauer M (2008) Molecular diagnosis of German patients with late-onset glycogen storage disease type II. J *Inherit. Metab. Dis.* doi:10.1007/s10545-008-0820-2.

[33] Bembi B, Cerini E, Danesino C, Donati MA, Gasperini S, Morandi L, Musumeci O, Parenti G, Ravaglia S, Seidita F, Toscano A, Vianello A (2008) Diagnosis of glycogenosis type II. *Neurology* 71 (23 Suppl 2):S4-11. doi:71/23_suppl_2/S4 [pii]. 10.1212/WNL.0b013e31818da91e.

[34] Winkel LP, Hagemans ML, van Doorn PA, Loonen MC, Hop WJ, Reuser AJ, van der Ploeg AT (2005) The natural course of non-classic Pompe's disease; a review of 225 published cases. *J. Neurol.* 252 (8):875-884. doi:10.1007/s00415-005-0922-9.

[35] Montalvo AL, Bembi B, Donnarumma M, Filocamo M, Parenti G, Rossi M, Merlini L, Buratti E, De Filippi P, Dardis A, Stroppiano M, Ciana G, Pittis MG (2006) Mutation profile of the GAA gene in 40 Italian patients with late onset glycogen storage disease type II. *Hum. Mutat.* 27 (10):999-1006. doi:10.1002/humu.20374.

[36] Leslie N, Tinkle BT (1993) Glycogen Storage Disease Type II (Pompe Disease). In: Pagon RA, Adam MP, Bird TD, Dolan CR, Fong CT, Stephens K (eds) *GeneReviews*. Seattle (WA).

[37] Vorgerd M, Burwinkel B, Reichmann H, Malin JP, Kilimann MW (1998) Adult-onset glycogen storage disease type II: phenotypic and allelic heterogeneity in German patients. *Neurogenetics* 1 (3):205-211.

[38] Pittis MG, Filocamo M (2007) Molecular genetics of late onset glycogen storage disease II in Italy. *Acta Myol.* 26 (1):67-71.

[39] Laforet P, Nicolino M, Eymard PB, Puech JP, Caillaud C, Poenaru L, Fardeau M (2000) Juvenile and adult-onset acid maltase deficiency in France: genotype-phenotype correlation. *Neurology* 55 (8):1122-1128.

[40] Remiche G, Ronchi D, Lamperti C, D'Angelo M, Bordoni A, Montrasio C, Crugnola V, Magri F, Moggio M, Bresolin N, Comi G (2008) Adult

Form Type II Glycogen Storage Disease in a Northern Italy Population: Phenotype Characterization, Early Diagnosis and Prognostic Determinants. *Neurology* 70 (11 Suppl 1):A309.

[41] Carlier RY, Laforet P, Wary C, Mompoint D, Laloui K, Pellegrini N, Annane D, Carlier PG, Orlikowski D (2011) Whole-body muscle MRI in 20 patients suffering from late onset Pompe disease: Involvement patterns. *Neuromuscul. Disord.* 21 (11):791-799. doi:S0960-8966 (11)00918-7 [pii]. 10.1016/j.nmd.2011.06.748.

[42] Remiche G, Herbaut AG, Ronchi D, Lamperti C, Magri F, Moggio M, Bresolin N, Comi GP (2012) Incontinence in late-onset Pompe disease: an underdiagnosed treatable condition. *Eur. Neurol.* 68 (2):75-78. doi:000338776 [pii]. 10.1159/000338776.

[43] van der Ploeg AT (2005) Monitoring of pulmonary function in Pompe disease: a muscle disease with new therapeutic perspectives. *Eur. Respir. J.* 26 (6):984-985. doi:26/6/984 [pii]. 10.1183/09031936.05.00112005.

[44] Remiche G, Lo Mauro A, Tarsia P, Ronchi D, Bordoni A, Magri F, Comi GP, Aliverti A, D'Angelo MG (2013) Postural effects on lung and chest wall volumes in late onset type II glycogenosis patients. *Respiratory physiology & neurobiology* 186 (3):308-314. doi:10.1016/ j.resp.2013.03.004.

[45] Pellegrini N, Laforet P, Orlikowski D, Pellegrini M, Caillaud C, Eymard B, Raphael JC, Lofaso F (2005) Respiratory insufficiency and limb muscle weakness in adults with Pompe's disease. *Eur. Respir. J.* 26 (6):1024-1031. doi:26/6/1024 [pii]. 10.1183/09031936.05.00020005.

[46] Rosenow EC, 3rd, Engel AG (1978) Acid maltase deficiency in adults presenting as respiratory failure. *The American journal of medicine* 64 (3):485-491.

[47] Keunen RW, Lambregts PC, Op de Coul AA, Joosten EM (1984) Respiratory failure as initial symptom of acid maltase deficiency. *Journal of neurology, neurosurgery, and psychiatry* 47 (5):549-552.

[48] Mellies U, Ragette R, Schwake C, Baethmann M, Voit T, Teschler H (2001) Sleep-disordered breathing and respiratory failure in acid maltase deficiency. *Neurology* 57 (7):1290-1295.

[49] Mellies U, Stehling F, Dohna-Schwake C, Ragette R, Teschler H, Voit T (2005) Respiratory failure in Pompe disease: treatment with noninvasive ventilation. *Neurology* 64 (8):1465-1467. doi:64/8/1465 [pii]. 10.1212/01.WNL.0000158682.85052.C0.

[50] Mellies U, Ragette R, Schwake C, Boehm H, Voit T, Teschler H (2003) Daytime predictors of sleep disordered breathing in children and

adolescents with neuromuscular disorders. *Neuromuscular disorders: NMD* 13 (2):123-128.

[51] Engel A, Franzini-Armstrong C (2004) *Myology : basic and clinical.* 3rd edn. McGraw-Hill, Medical Pub. Division, New York.

[52] Wokke JH, Escolar DM, Pestronk A, Jaffe KM, Carter GT, van den Berg LH, Florence JM, Mayhew J, Skrinar A, Corzo D, Laforet P (2008) Clinical features of late-onset Pompe disease: a prospective cohort study. *Muscle Nerve* 38 (4):1236-1245. doi:10.1002/mus.21025.

[53] van der Beek NA, van Capelle CI, van der Velden-van Etten KI, Hop WC, van den Berg B, Reuser AJ, van Doorn PA, van der Ploeg AT, Stam H (2011) Rate of progression and predictive factors for pulmonary outcome in children and adults with Pompe disease. *Mol. Genet. Metab.* 104 (1-2):129-136. doi:S1096-7192(11)00199-5 [pii]. 10.1016/j.ymgme. 2011.06.012.

[54] Prigent H, Orlikowski D, Laforet P, Letilly N, Falaize L, Pellegrini N, Annane D, Raphael JC, Lofaso F (2012) Supine volume drop and diaphragmatic function in adults with Pompe disease. *Eur. Respir. J.* 39 (6):1545-1546. doi:39/6/1545 [pii]. 10.1183/09031936.00169011.

[55] Fromageot C, Lofaso F, Annane D, Falaize L, Lejaille M, Clair B, Gajdos P, Raphael JC (2001) Supine fall in lung volumes in the assessment of diaphragmatic weakness in neuromuscular disorders. Arch *Phys. Med. Rehabil.* 82 (1):123-128. doi:S0003-9993(01)33957-6 [pii]. 10.1053/apmr.2001.18053.

[56] Bijvoet AG, Van Hirtum H, Vermey M, Van Leenen D, Van Der Ploeg AT, Mooi WJ, Reuser AJ (1999) Pathological features of glycogen storage disease type II highlighted in the knockout mouse model. *J. Pathol.* 189 (3):416-424. doi:10.1002/(SICI)1096-9896(199911)189: 3<416::AID-PATH445>3.0.CO;2-6 [pii]. 10.1002/(SICI)1096-9896 (199911)189:3<416::AID-PATH445>3.0.CO;2-6.

[57] Teng YT, Su WJ, Hou JW, Huang SF (2004) Infantile-onset glycogen storage disease type II (Pompe disease): report of a case with genetic diagnosis and pathological findings. *Chang Gung Med. J.* 27 (5):379-384. doi:2705/270508 [pii].

[58] Thurberg BL, Lynch Maloney C, Vaccaro C, Afonso K, Tsai AC, Bossen E, Kishnani PS, O'Callaghan M (2006) Characterization of pre- and post-treatment pathology after enzyme replacement therapy for Pompe disease. *Lab. Invest.* 86 (12):1208-1220. doi:3700484 [pii]. 10.1038/labinvest.3700484.

[59] Schuller A, Wenninger S, Strigl-Pill N, Schoser B (2012) Toward deconstructing the phenotype of late-onset Pompe disease. *Am. J. Med. Genet. C Semin. Med. Genet.* 160 (1):80-88. doi:10.1002/ajmg.c.31322.

[60] Schoser BG, Muller-Hocker J, Horvath R, Gempel K, Pongratz D, Lochmuller H, Muller-Felber W (2007) Adult-onset glycogen storage disease type 2: clinico-pathological phenotype revisited. *Neuropathol. Appl. Neurobiol.* 33 (5):544-559. doi:NAN839 [pii]. 10.1111/j.1365-2990.2007.00839.x.

[61] Sacconi S, Bocquet JD, Chanalet S, Tanant V, Salviati L, Desnuelle C (2010) Abnormalities of cerebral arteries are frequent in patients with late-onset Pompe disease. *J. Neurol.* doi:10.1007/s00415-010-5618-0.

[62] Laforet P, Petiot P, Nicolino M, Orlikowski D, Caillaud C, Pellegrini N, Froissart R, Petitjean T, Maire I, Chabriat H, Hadrane L, Annane D, Eymard B (2008) Dilative arteriopathy and basilar artery dolichoectasia complicating late-onset Pompe disease. *Neurology* 70 (22):2063-2066. doi:70/22/2063 [pii]. 10.1212/01.wnl.0000313367.09469.13.

[63] Patel TT, Banugaria SG, Frush DP, Enterline DS, Tanpaiboon P, Kishnani PS (2013) Basilar artery aneurysm: a new finding in classic infantile Pompe disease. *Muscle & nerve* 47 (4):613-615. doi:10.1002/mus.23659.

[64] Felice KJ, Alessi AG, Grunnet ML (1995) Clinical variability in adult-onset acid maltase deficiency: report of affected sibs and review of the literature. *Medicine* (Baltimore) 74 (3):131-135.

[65] Kroos M, Manta P, Mavridou I, Muntoni F, Halley D, Van der Helm R, Zaifeiriou D, Van der Ploeg A, Reuser A, Michelakakis H (2006) Seven cases of Pompe disease from Greece. *J. Inherit. Metab. Dis.* 29 (4):556-563. doi:10.1007/s10545-006-0280-5.

[66] Groen WB, Leen WG, Vos AM, Cruysberg JR, van Doorn PA, van Engelen BG (2006) Ptosis as a feature of late-onset glycogenosis type II. *Neurology* 67 (12):2261-2262. doi:10.1212/01.wnl.0000249183.39952.3e.

[67] Case LE, Beckemeyer AA, Kishnani PS (2012) Infantile Pompe disease on ERT: update on clinical presentation, musculoskeletal management, and exercise considerations. American journal of medical genetics Part C, *Seminars in medical genetics* 160 (1):69-79. doi:10.1002/ajmg.c.31321.

[68] van Gelder CM, van Capelle CI, Ebbink BJ, Moor-van Nugteren I, van den Hout JM, Hakkesteegt MM, van Doorn PA, de Coo IF, Reuser AJ, de Gier HH, van der Ploeg AT (2012) Facial-muscle weakness, speech

disorders and dysphagia are common in patients with classic infantile Pompe disease treated with enzyme therapy. *Journal of inherited metabolic disease* 35 (3):505-511. doi:10.1007/s10545-011-9404-7.

[69] Prater SN, Banugaria SG, DeArmey SM, Botha EG, Stege EM, Case LE, Jones HN, Phornphutkul C, Wang RY, Young SP, Kishnani PS (2012) The emerging phenotype of long-term survivors with infantile Pompe disease. *Genetics in medicine : Official Journal of the American College of Medical Genetics* 14 (9):800-810. doi:10.1038/gim.2012.44.

[70] Dubrovsky A, Corderi J, Lin M, Kishnani PS, Jones HN (2011) Expanding the phenotype of late-onset Pompe disease: tongue weakness: a new clinical observation. *Muscle & nerve* 44 (6):897-901. doi:10.1002/mus.22202.

[71] Kamphoven JH, de Ruiter MM, Winkel LP, Van den Hout HM, Bijman J, De Zeeuw CI, Hoeve HL, Van Zanten BA, Van der Ploeg AT, Reuser AJ (2004) Hearing loss in infantile Pompe's disease and determination of underlying pathology in the knockout mouse. *Neurobiol. Dis.* 16 (1):14-20. doi:10.1016/j.nbd.2003.12.018.

[72] van Capelle CI, Goedegebure A, Homans NC, Hoeve HL, Reuser AJ, van der Ploeg AT (2010) Hearing loss in Pompe disease revisited: results from a study of 24 children. *Journal of inherited metabolic disease* 33 (5):597-602. doi:10.1007/s10545-010-9144-0.

[73] Musumeci O, Catalano N, Barca E, Ravaglia S, Fiumara A, Gangemi G, Rodolico C, Sorge G, Vita G, Galletti F, Toscano A (2012) Auditory system involvement in late onset Pompe disease: a study of 20 Italian patients. *Mol. Genet. Metab.* 107 (3):480-484. doi:10.1016/j.ymgme.2012.07.024. S1096-7192(12)00299-5 [pii]

[74] van der Beek NA, Verschuure H, Reuser AJ, van der Ploeg AT, van Doorn PA, Poublon RM (2012) Hearing in adults with Pompe disease. *Journal of inherited metabolic disease* 35 (2):335-341. doi:10.1007/s10545-011-9396-3.

[75] Tan QK, Cheah SM, Dearmey SM, Kishnani PS (2013) Low anal sphincter tone in infantile-onset Pompe Disease: an emerging clinical issue in enzyme replacement therapy patients requiring special attention. *Molecular genetics and metabolism* 108 (2):142-144. doi:10.1016/j.ymgme.2012.11.013.

[76] Remiche G.; Ronchi, D., Lamperti, C.; Bordoni, A.; Magri, F.; Moggio, M.; Comi, GP. (2013) Spontaneous hydromyelic cavity in two unrelated patients with late onset Pompe disease: is this a fortuitous association? . *Eur. Neurol.* 70 (1-2). doi:10.1159/000350851.

[77] Hobson-Webb LD, Proia AD, Thurberg BL, Banugaria S, Prater SN, Kishnani PS (2012) Autopsy findings in late-onset Pompe disease: A case report and systematic review of the literature. *Mol. Genet. Metab.* 106 (4):462-469. doi:S1096-7192(12)00196-5 [pii]. 10.1016/j.ymgme. 2012.05.007.

[78] Wary C, Laforet P, Eymard B, Fardeau M, Leroy-Willig A, Bassez G, Leroy JP, Caillaud C, Poenaru L, Carlier PG (2003) Evaluation of muscle glycogen content by 13C NMR spectroscopy in adult-onset acid maltase deficiency. *Neuromuscul. Disord.* 13 (7-8):545-553. doi: S0960896603000695 [pii].

[79] Bembi B, Pisa FE, Confalonieri M, Ciana G, Fiumara A, Parini R, Rigoldi M, Moglia A, Costa A, Carlucci A, Danesino C, Pittis MG, Dardis A, Ravaglia S (2010) Long-term observational, non-randomized study of enzyme replacement therapy in late-onset glycogenosis type II. *J. Inherit. Metab. Dis.* 33 (6):727-735. doi:10.1007/s10545-010-9201-8.

[80] Wokke JH, Ausems MG, van den Boogaard MJ, Ippel EF, van Diggelene O, Kroos MA, Boer M, Jennekens FG, Reuser AJ, Ploos van Amstel HK (1995) Genotype-phenotype correlation in adult-onset acid maltase deficiency. *Ann. Neurol.* 38 (3):450-454. doi:10.1002/ana. 410380316.

[81] Muller-Felber W, Horvath R, Gempel K, Podskarbi T, Shin Y, Pongratz D, Walter MC, Baethmann M, Schlotter-Weigel B, Lochmuller H, Schoser B (2007) Late onset Pompe disease: clinical and neurophysiological spectrum of 38 patients including long-term follow-up in 18 patients. *Neuromuscul. Disord.* 17 (9-10):698-706. doi:S0960-8966(07)00187-3 [pii]. 10.1016/j.nmd.2007.06.002.

[82] Spada M, Porta F, Vercelli L, Pagliardini V, Chiado-Piat L, Boffi P, Pagliardini S, Remiche G, Ronchi D, Comi G, Mongini T (2013) Screening for later-onset Pompe's disease in patients with paucisymptomatic hyperCKemia. *Molecular genetics and metabolism* 109 (2):171-173. doi:10.1016/j.ymgme.2013.03.002.

[83] Li Y, Scott CR, Chamoles NA, Ghavami A, Pinto BM, Turecek F, Gelb MH (2004) Direct multiplex assay of lysosomal enzymes in dried blood spots for newborn screening. *Clin. Chem.* 50 (10):1785-1796. doi: 10.1373/clinchem.2004.035907. clinchem.2004.035907 [pii].

[84] Winchester B, Bali D, Bodamer OA, Caillaud C, Christensen E, Cooper A, Cupler E, Deschauer M, Fumic K, Jackson M, Kishnani P, Lacerda L, Ledvinova J, Lugowska A, Lukacs Z, Maire I, Mandel H, Mengel E, Muller-Felber W, Piraud M, Reuser A, Rupar T, Sinigerska I, Szlago M,

Verheijen F, van Diggelen OP, Wuyts B, Zakharova E, Keutzer J (2008) Methods for a prompt and reliable laboratory diagnosis of Pompe disease: report from an international consensus meeting. *Mol. Genet. Metab.* 93 (3):275-281. doi:S1096-7192(07)00407-6 [pii]. 10.1016/j.ymgme.2007.09.006.

[85] Kishnani PS, Steiner RD, Bali D, Berger K, Byrne BJ, Case LE, Crowley JF, Downs S, Howell RR, Kravitz RM, Mackey J, Marsden D, Martins AM, Millington DS, Nicolino M, O'Grady G, Patterson MC, Rapoport DM, Slonim A, Spencer CT, Tifft CJ, Watson MS (2006) Pompe disease diagnosis and management guideline. *Genet. Med.* 8 (5):267-288. doi:10.109701.gim.0000218152.87434.f3. 00125817-200605000-00001 [pii].

[86] Kallwass H, Carr C, Gerrein J, Titlow M, Pomponio R, Bali D, Dai J, Kishnani P, Skrinar A, Corzo D, Keutzer J (2007) Rapid diagnosis of late-onset Pompe disease by fluorometric assay of alpha-glucosidase activities in dried blood spots. *Mol. Genet. Metab.* 90 (4):449-452. doi:S1096-7192(06)00389-1 [pii]. 10.1016/j.ymgme.2006.12.006.

[87] Diagnostic criteria for late-onset (childhood and adult) Pompe disease (2009). *Muscle Nerve* 40 (1):149-160. doi:10.1002/mus.21393.

[88] Goldstein JL, Young SP, Changela M, Dickerson GH, Zhang H, Dai J, Peterson D, Millington DS, Kishnani PS, Bali DS (2009) Screening for Pompe disease using a rapid dried blood spot method: experience of a clinical diagnostic laboratory. *Muscle Nerve* 40 (1):32-36. doi:10.1002/mus.21376

[89] Chien YH, Chiang SC, Zhang XK, Keutzer J, Lee NC, Huang AC, Chen CA, Wu MH, Huang PH, Tsai FJ, Chen YT, Hwu WL (2008) Early detection of Pompe disease by newborn screening is feasible: results from the Taiwan screening program. *Pediatrics* 122 (1):e39-45. doi:peds.2007-2222 [pii]. 10.1542/peds.2007-2222.

[90] Desnuelle C, Salviati L (2011) Challenges in diagnosis and treatment of late-onset Pompe disease. *Curr. Opin. Neurol.* 24 (5):443-448. doi: 10.1097/WCO.0b013e32834a1e00.

[91] Hobson-Webb LD, Dearmey S, Kishnani PS (2011) The clinical and electrodiagnostic characteristics of Pompe disease with post-enzyme replacement therapy findings. *Clinical neurophysiology: Official Journal of the International Federation of Clinical Neurophysiology* 122 (11):2312-2317. doi:10.1016/j.clinph.2011.04.016.

[92] Dubowitz V, Sewry CA, Lane RJM (2007) *Muscle biopsy: a practical approach.* 3rd edn. Saunders Elsevier, Philadelphia.

[93] Ausems MG, Lochman P, van Diggelen OP, Ploos van Amstel HK, Reuser AJ, Wokke JH (1999) A diagnostic protocol for adult-onset glycogen storage disease type II. *Neurology* 52 (4):851-853.

[94] Hermans MM, van Leenen D, Kroos MA, Beesley CE, Van Der Ploeg AT, Sakuraba H, Wevers R, Kleijer W, Michelakakis H, Kirk EP, Fletcher J, Bosshard N, Basel-Vanagaite L, Besley G, Reuser AJ (2004) Twenty-two novel mutations in the lysosomal alpha-glucosidase gene (GAA) underscore the genotype-phenotype correlation in glycogen storage disease type II. *Hum. Mutat.* 23 (1):47-56. doi:10.1002/humu.10286.

[95] Laloui K, Wary C, Carlier RY, Hogrel JY, Caillaud C, Laforet P (2011) Making diagnosis of Pompe disease at a presymptomatic stage: to treat or not to treat? *Neurology* 77 (6):594-595. doi:WNL.0b013e318228c0ea [pii]. 10.1212/WNL.0b013e318228c0ea.

[96] Jones LK, Jr., Liewluck T, Gavrilova RH (2012) Myalgic phenotype and preserved muscle strength in adult-onset acid maltase deficiency. *Neuromuscular disorders*: NMD 22 (8):763-766. doi:10.1016/j.nmd.2012.02.003.

[97] Kroos MA, Waitfield AE, Joosse M, Winchester B, Reuser AJ, MacDermot KD (1997) A novel acid alpha-glucosidase mutation identified in a Pakistani family with glycogen storage disease type II. *Journal of inherited metabolic disease* 20 (4):556-558.

[98] Van der Kraan M, Kroos MA, Joosse M, Bijvoet AG, Verbeet MP, Kleijer WJ, Reuser AJ (1994) Deletion of exon 18 is a frequent mutation in glycogen storage disease type II. *Biochemical and biophysical research communications* 203 (3):1535-1541.

[99] Lee HS, Daniels BH, Salas E, Bollen AW, Debnath J, Margeta M (2012) Clinical utility of LC3 and p62 immunohistochemistry in diagnosis of drug-induced autophagic vacuolar myopathies: a case-control study. *PLoS One* 7 (4):e36221. doi:10.1371/journal.pone.0036221.

[100] Beydon N, Davis SD, Lombardi E, Allen JL, Arets HG, Aurora P, Bisgaard H, Davis GM, Ducharme FM, Eigen H, Gappa M, Gaultier C, Gustafsson PM, Hall GL, Hantos Z, Healy MJ, Jones MH, Klug B, Lodrup Carlsen KC, McKenzie SA, Marchal F, Mayer OH, Merkus PJ, Morris MG, Oostveen E, Pillow JJ, Seddon PC, Silverman M, Sly PD, Stocks J, Tepper RS, Vilozni D, Wilson NM (2007) An official American Thoracic Society/European Respiratory Society statement: pulmonary function testing in preschool children. *American Journal of*

Enzyme replacement therapy with alglucosidase alfa in 44 patients with late-onset glycogen storage disease type 2: 12-month results of an observational clinical trial. *J. Neurol.* 257 (1):91-97. doi:10.1007/s00415-009-5275-3.

[117] Cartier N, Hacein-Bey-Abina S, Bartholomae CC, Bougneres P, Schmidt M, Kalle CV, Fischer A, Cavazzana-Calvo M, Aubourg P (2012) Lentiviral hematopoietic cell gene therapy for X-linked adrenoleukodystrophy. *Methods in enzymology* 507:187-198. doi:10.1016/B978-0-12-386509-0.00010-7.

[118] Booth C, Gaspar HB, Thrasher AJ (2011) Gene therapy for primary immunodeficiency. *Curr. Opin. Pediatr.* 23 (6):659-666. doi:10.1097/MOP.0b013e32834cd67a.

[119] Jones HN, Muller CW, Lin M, Banugaria SG, Case LE, Li JS, O'Grady G, Heller JH, Kishnani PS (2010) Oropharyngeal dysphagia in infants and children with infantile Pompe disease. *Dysphagia* 25 (4):277-283. doi:10.1007/s00455-009-9252-x.

[120] Hobson-Webb LD, Jones HN, Kishnani PS (2013) Oropharyngeal dysphagia may occur in late-onset Pompe disease, implicating bulbar muscle involvement. *Neuromuscular disorders*: NMD 23 (4):319-323. doi:10.1016/j.nmd.2012.12.003.

In: Glycogen
Editors: P.L. Weiss and B.D. Faulkner
ISBN: 978-1-62948-395-5

Chapter 2

Glycogen Metabolism during Chronic Liver Diseases

***Natalia N. Bezborodkina*[*], *Anna Yu. Chestnova*[†] *and Boris N. Kudryavtsev*[‡]**

Laboratory of Cellular Pathology, Institute of Cytology of the Russian Academy of Sciences, Saint Petersburg, Russia

Abstract

The chapter focuses on the metabolism of glycogen and glucose during chronic hepatitides and cirrhosis of various etiology. Chronic hepatitides and their final stage, liver cirrhosis, are common and severe human diseases. They cause the impairment of the liver's morphological structure as well as various changes in the metabolism of carbohydrates, including that of glycogen, which is reflected in the glucostatic function of the liver. In particular, carbohydrate metabolism during cirrhosis is similar to that during diabetes, whose main feature is tissue intolerance to glucose. The chapter mostly presents the results of the authors' original research. Carbohydrate metabolism in the normal and the pathologically changed liver is examined during starvation and at various stages of the

[*] Corresponding author N. N. Bezborodkina, Ph.D. Laboratory of Cellular Pathology, Institute of Cytology of the Russian Academy of Sciences, Tikhoretsky ave 4, 194064 Saint Petersburg, Russia. E-mail: natalia_bezborodkina@mail.ru.

[†] E-mail: chestnovaanna@mail.ru

[‡] E-mail: bn_kudryavtsev@mail.ru.

trophic cycle. The chapter contains a detailed discussion of regulatory mechanisms of glycogen metabolism in the normal and the cirrhotic liver, which help maintain the glucose homeostasis during starvation, in the absorptive and post-absorptive periods of the trophic cycle. Data on the activity of several key enzymes of glycolysis, gluconeogenesis, glycogenesis and glycogenolysis are given. Much attention is given to the glucose and glycogen metabolism under conditions of changing severity of the liver's chronic impairment.

INTRODUCTION

Cirrhosis, the final stage of chronic hepatitis, is characterised by structural impairment of liver and dramatic changes in hepatic carbohydrate metabolism, including that of glucose and glycogen. Cirrhotic patients often show decreased glucose tolerance after a meal [1, 2]. An elevated glucose level in the blood of such patients is accompanied by increased concentrations of lactate, pyruvate and insulin [3, 4].

Mechanisms of glucose level elevation in the blood during liver diseases are still obscure. It was suggested that glucose intolerance may be due to the decreased glucose absorption in both the liver and the peripheral tissues. Possible factors conducive to glucose intolerance are malnutrition, decreased weight of liver parenchyma, insensitivity of tissues to insulin and impaired hepatic glycogen synthesis [2].

Information about glycogen content in pathologically changed liver is scarce and rather contradictory. Patients with acute viral hepatitis were shown to have lower glycogen concentrations in the liver [5]. In patients with alcoholic cirrhosis glycogen content in the liver made up 25.8±3.5 mg/g of wet weight, that is, less than in healthy humans (32.0-48.8 mg/g of wet weight) [6]. Besides, glycogen concentration in the liver of cirrhotic patients 5-6 hours after a meal was much lower than in healthy volunteers [7]. Glycogen content in the liver of cirrhotic rats was 64% lower than in the control [8]. However, in some other studies glycogen content in the cirrhotic liver was within the norm or even higher than the norm [2, 9, 10]. As shown by cytophotometric methods, glycogen content in hepatocytes isolated from the liver material biopsied from patients with chronic hepatitis and cirrhosis of various etiology exceeded the norm several times [11].

Research on animals has demonstrated an intimate connection between glycogen content in the liver and the direction of carbon flow through glycolysis or gluconeogenesis [12, 13]. Though some authors think that the rate of gluconeogenesis in cirrhotic liver is the same as in the normal one or even lower [14], most of the data point to an enhanced rate [8, 15, 16]. The contribution of glycogenolysis and gluconeogenesis into glucose production changes as a result. In the normal liver after nocturnal fasting 70-80% of glucose is formed by glycogenolysis and the rest, by gluconeogenesis [2, 17]. On the contrary, in cirrhotic liver gluconeogenesis accounts for 67-87% of glucose production [7, 8].

Glucose concentration in the blood during fasting reflects the balance between its production in the liver and its assimilation in the peripheral tissues. If the rates of these processes rise or fall simultaneously, the turnover of glucose changes but its level in the blood does not. Therefore, a normal glucose level in the blood of patients with chronic liver diseases [1, 8] does not mean that hepatic glucose production is also normal. In this connection, it is important to understand the mechanisms maintaining glucostatic function in the cirrhotic liver during fasting and in the absorptive and post-absorptive period (after meals).

Glucose and Glycogen Metabolism in the Liver during Fasting

Hepatic carbohydrate metabolism changes considerably during fasting, especially if it is prolonged [18, 19]. Glycogen reserves in the liver, normally sufficient for ca. 24 h, are depleted, and so are protein reserves in muscular tissues, the source of amino-acids for hepatic gluconeogenesis. The organism is gradually adjusting to lower glucose availability. For instance, glucose requirements of brain decrease almost three-fold during starvation, being mostly ensured by oxidation of ketone bodies formed in the liver by incomplete oxidation of fatty acids [2, 20].

In our research, glycogen content in the normal and the cirrhotic rat liver was at a very low but approximately the same level after 48 h of fasting (Table 1). This result is in good agreement with the reports of other authors obtained on rats fasted for 24 and 48 h [8, 21].

Glycogen content in hepatic cells is determined by the rates of its synthesis and degradation. These processes involve numerous enzymes, whose activity is influenced, among other things, by glucose concentration in the blood. Contrary to peripheral tissues, glucose transport across the membrane of hepatic cells is not insulin-mediated, proceeding, at a high rate, by facilitated diffusion. Therefore, even slight fluctuations in blood glucose may have an immediate and dramatic effect on the activity of the enzymes involved in glycogen metabolism, first of all, glucokinase, glycogen synthase and glycogen phosphorylase.

Table 1. Several data of carbohydrate metabolism in blood and liver of starving rats in norm and at the cirrhosis (values are means with their standard errors for three rats)

	normal	cirrhosis
Glucose, mmol/l	3.24±0.28	3.16±0.14
Liver glycogen, µmol glucosyl units/ g ww of liver	15.4±0.2	10.4±0.5
Glucokinase, nmol/min per 1 mg of protein	11.2±0.2	1.9±0.1
Glucose-6-phosphatase, nmol/min per 1 mg of protein	44.2±0.6	43.4±1.3
Phosphofructokinase, nmol/min per 1 mg of protein	8.9±1.0	27.9±1.4
Fructose-1,6-diphosphatase, nmol/min per 1 mg of protein	3.3±0.1	9.7±0.7
Glycogen synthase, -G-6-P/+G-6-P	0.19±0.01	0.20±0.01
Glycogen phosphorylase (active form), nmol/min per 1 mg of protein	99.0±2.4	44.1±1.6
Glycogen phosphorylase (total), nmol/min per 1 mg of protein	162.4±3.1	158.9±1.5

In our research, glucose concentration in the peripheral blood of the normal and the cirrhotic rats after 48 h of fasting was at a low level, making up 3.2 mM/L (Table 1). Similar values (3.0–3.2 mM/L) in the blood of rats fasted for 40-48 h were reported by other authors [22, 23, 24].

Glucokinase (GK), the key enzyme responsible for glucose transport to hepatic cells at the start of glycolysis and glycogenesis, is normally in excess as compared its substrate, glucose. GK amount and activity in the liver is glucose- as well as insulin-controlled [25, 26]. GK activity falls during fasting and increases in replete state proportionally to the increase in glucose concentration. Since GK K_m makes up ca. 20 mM [27], the rate of glucose phosphorylation under physiological conditions is determined only by its concentration in the cell. Therefore, low blood glucose, such as is observed during fasting, may be expected to inhibit GK and to channel the substrate flow in the liver towards glucose production. In this research, GK activity in the normal liver of fasted rats was decreased when blood glucose was low. This result agrees with the literature [28, 29]. During prolonged fasting GK activity continues to decrease [29].

The activity of phosphofructokinase (PFK) is known to decrease with the decreasing concentrations of hepatic hexose monophosphates, in particular, fructose-6-phosphates, at late stages of fasting, which is due to depletion of glycogen reserves [28, 30]. Low PFK activity in the liver of rats fasted for 48 h reported in this chapter supports the above observations. Low activities of both PFK and GK indicate a low intensity of glycolysis during prolonged fasting.

The main function of glycolysis in the liver is storage of precursors for the biosynthesis of fatty acids and, to a lesser extent, formation of pyruvate for oxidative processes. However, glycolysis is unlikely to play an important role in these processes during fasting, given its low rate in the liver of fasted rats. During fasting, the cells tap much of their energy requirements from β-oxidation of fatty acids essential for the ketone bodies formation [2, 20].

Contrary to GK and PFK, the activity of glucose-6-phosphatase (G6Pase) and fructose-1,6-diphospatase (Fru-1,6-Pase) in the liver of fasted rats was at a high level (Table 1). This result is in agreement with the idea that when glycogen is depleted, gluconeogenesis and ketogenesis step forward as the key processes in the liver [19, 20].

Low concentrations of glucose, increasing concentrations of glucagon and decreasing concentrations of insulin in the blood during fasting promote transformation of glycogen phosphorylase (GP) from its b-form into a-form as well as glycogen synthase (GS) inactivation [31, 32]. Our results (Table 1) support the idea that glycogen metabolism in the liver of fasted rats is characterised by low GS activity and high GP*a* activity [33, 34, 35].

Active GP form is known to be formed with participation of kinase phosphorylase, which, in turn, is activated by cAMP-dependent protein kinase A [36, 37]. The concentration of cAMP, whose synthesis is stimulated by glucagon, increases by 60% in the liver of fasted rats [38]. This fact lends further support to the idea that metabolic conditions in the liver of fasted animals are conducive to transformation of GP into active GP*a* and to enhanced glycogenolysis.

Though blood glucose and hepatic glycogen in the cirrhotic rats after 48 h of fasting were the same as in the control animals, carbohydrate metabolism in the cirrhotic liver was different. Most available information about carbohydrate metabolism in the cirrhotic liver during fasting was obtained on humans after nocturnal fasting (8-12 h). In such cases, the liver was shown to produce glucose both by gluconeogenesis and by glycogenolysis. In contrast to nocturnal fasting, glucose production by cirrhotic rat liver after 48 h of fasting proceeded only by gluconeogenesis, as glycogen reserves were very low. The fact that blood glucose in cirrhotic humans or rats is normal or slightly elevated during fasting does not necessarily mean the hepatic glucose production is normal, too. In several studies, hepatic glucose production after nocturnal fasting was 20-40% lower during cirrhosis as compared to the norm [24, 39]. In other studies, however, glucose production by the cirrhotic liver was normal, even though it was sustained owing to increased glucagon concentrations [40, 41]. These contradictions concerning glucose production in the cirrhotic liver could be resolved by determining the activity of key gluconeogenesis enzymes.

G6Pase plays a key role in hepatic glucose production. This enzyme splits the phosphate group from glucose-6-phosphate formed during glycogenolysis or gluconeogenesis, and free glucose is released into blood as a result [42]. G6Pase activity in cirrhotic liver of rats fasted for 48 h was the same as in the norm (Table 1). This means that glucose production in the cirrhotic liver of fasted rats, which proceeds only by gluconeogenesis, lies within the norm.

The activity of other key gluconeogenesis enzymes – pyruvate carboxylase, phosphoenolpyruvate carboxykinase and Fru-1,6-Pase – in the liver of patients with alcoholic cirrhosis after nocturnal fasting was normal, when calculated per liver weight unit. It should be noted, however, that the amount of connective tissue in the cirrhotic liver is 4-7 times greater than in the normal one [43, 44, 45]. Therefore, "normal" or "decreased" activity values of gluconeogenesis enzymes in cirrhotic liver may not reflect the actual state of things. If calculated per weight unit of viable hepatic parenchyma, they may well turn out to be more active than in the norm. Our results concerning Fru-1,6-Pase activity in the liver of fasted rats support this conjecture. It was 3 times higher in the cirrhotic liver as compared to the normal one (Table 1). This result may indicate an increased rate of gluconeogenesis in the rat liver after 48 h of fasting (Table 1). Indeed, though some authors report normal or even lower gluconeogenesis rate during cirrhosis [14], most available evidence indicates that its rate is increased [7, 15, 16].

If gluconeogenesis rate is increased during cirrhosis, then, by inference, this process should be more intensively supplied by substrates such as lactate, pyruvate, glycerol and some amino-acids. The most important substrates during prolonged fasting are amino-acids (alanine, glutamine and glycine), which are formed during protein breakdown in skeletal muscles, and glycerol, which is formed by utilization of triglycerides transported to the liver. Little is known about the production of gluconeogenic precursors by peripheral tissues during liver cirrhosis. The rate of glycerol production in patients with alcoholic liver diseases was shown to be normal [17]; evidence about other substrates is simply lacking. Increased concentrations of lactate, pyruvate and glycerol, often recorded in the blood of cirrhotic patients, are, therefore, not very informative, as they may possibly reflect a poorer clearance of these substances due to impairment of liver blood supply.

Though blood glucose in control and experimental rats after 48 h of fasting was the same, GK activity in the cirrhotic liver made up as little as 16% of the norm (Table 1). A similar abrupt fall of GK activity during biliary and alcoholic liver cirrhosis in humans has often been reported [27, 46, 47].

Glucose intolerance is a characteristic accompaniment of liver cirrhosis [40, 41, 48], which shares this feature with diabetes. GK activity decreases considerably during both these diseases [49, 50, 51]. As GK is important for glucose absorption by the liver, its decreased activity is thought to contribute to hyperglycemia [50, 51, 52]. Glucose intolerance in different tissues is generally considered to develop through impaired insulin secretion by the pancreas and/or cell insensitivity to this hormone. The latter may result from impaired binding of insulin with its receptors on the cell surface or from a change in the lipid composition of plasma membranes. As GK level in the liver is regulated by the level of its expression [53, 54] and its compartmentalization in the cell [55], which are insulin-mediated, impairment of insulin secretion or binding may result in an abrupt fall of GK activity during cirrhosis.

While GK activity in the cirrhotic liver of fasted rats was decreased, the activity of another key glycolysis enzyme, PFK, was increased ca. 2.8-fold (Table 1). This means that either glucose can be phosphorylated in the cirrhotic liver by some other enzyme than GK or that there is a substrate cycle at the level of PFK/Fru-1,6-Pase. The activity of glucose phosphorylating enzymes – hexokinase I and hexokinase II – in the cirrhotic liver is increased, respectively, 3- and 7-fold [27]. These data indicate that the fall of GK phosphorylating activity during cirrhosis may be partly compensated by an increase in the hexokinase activity. However, the exact contribution of hexokinases into increasing the rate of G6P (and, thus, F6P) formation remains unknown as there is no information about the concentration of G6P in the cirrhotic liver. Taking into account a much lower absolute activity of hexokinases as compared to GK, we cannot expect any considerable influence of hexokinases on increasing the substrate flow through PFK during cirrhosis.

Glycogen synthase (GS) activity in cirrhotic rat liver after 48 h of fasting was the same as in the normal liver (Table 1). Such low activity is probably due to the low blood glucose in fasted rats. Contrary to our research, Giardina et al. [21] showed that GS activity in the liver of rats with thioacetamide cirrhosis after 24 h of fasting was much lower than the norm. Similar data were obtained on rats with liver cirrhosis induced by bile duct ligation. A considerable fall of the total and active GS was shown to correlate with the decrease in the number of hepatic cells [56]. On the other hand, in the liver of rats with CCl_4-induced cirrhosis the total GS level increased and the level of its active form remained within the norm [8].

In our research, GP*a* activity in the cirrhotic liver of rats fasted for 48 h was decreased by more than 50% as compared to the norm (Table 1). GP activation is known to be cAMP-dependent [32, 57]. Adenylate cyclase activity and cAMP level decrease in the cirrhotic liver [58, 59], which, alongside with the decreased synthesis rate of this enzyme, probably plays a considerable role in the fall of GP*a* levels during this disease.

Absorptive Period

Nutrition is crucial for many aspects of animal welfare such as health, physical activity, reproductive capacity, fertility, adaptability and life span. Food uptake, processing and absorption in the intestine, as well as transport of nutrients to tissues and their utilisation in the cells are vitally important life processes.

Glucose and fructose make up much of the carbohydrates daily consumed by animals and humans with the food. Administration of these monosaccharides to fasted rats resulted in a rapid rise of blood glucose (Figure 1), which doubled 10 min after refeeding both in the experiment and in the control. Later blood glucose continued to increase but remained lower in the cirrhotic rat liver than in the control at all times after refeeding by glucose or fructose. This may be due both to a higher rate of glucose utilisation in the tissues and a lower glucose uptake from the intestine.

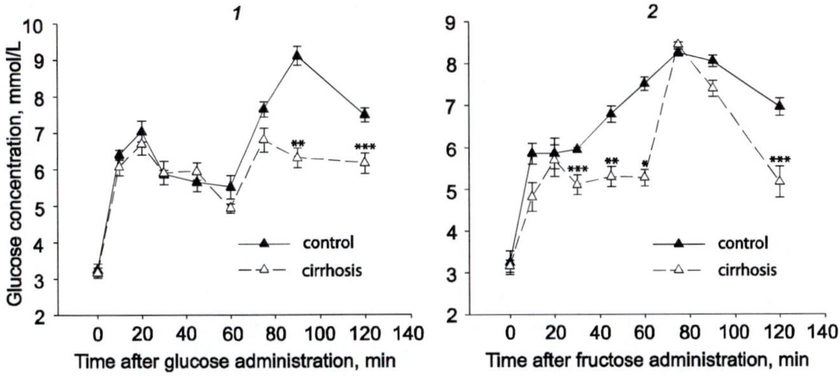

Figure 1. Blood serum glucose in normal and cirrhotic rats are different time intervals after refeeding with glucose (1) or fructose (2). *P < 0.001, **P < 0.01, ***P<0.05 – mean values were significantly different from control.

Glucose intolerance of various tissues often occurs in cirrhotic patients [1, 2, 10]. In such patients, absorption and utilisation rates of glucose contained in the food are extremely unlikely to be high. On the other hand, absorption and transport of sugars, amino-acids and some other substances diminished considerably during liver cirrhosis, as shown by research focussed on the intestinal condition during this disease [60]. Moreover, an impairment of intestinal function during liver cirrhosis is accompanied by structural changes of the intestine [61, 62, 63]. Therefore, a very probably reason of the slower glucose increase in the peripheral blood of rats with CCl_4-induced liver cirrhosis are structural-functional intestinal disorders.

The increase in blood glucose after glucose administration to fasted rats resulted in a rapid accumulation of glycogen in the normal liver. Its content doubled 10 min after refeeding and proceeded at a high rate, reaching the maximum after 90 min; after that it decreased (Figure 2).

Changes in glycogen content in normal liver after peroral glucose administration to fasted rats may be divided into three phases: 1) *glycogen synthesis* (0-4 h), 2) *stable phase* (4-12 h) and 3) *glycogen degradation* (12-24 h) [64]. This means that the decreased glycogen content 90 min after refeeding, reported in this chapter, probably does not correspond to the end of the glycogen synthesis phase but reflects the wavelike nature of this process.

High rate of glycogen synthesis in the liver after refeeding with monosaccharides has often been recorded. The rate of glycogen synthesis in the liver after peroral administration of glucose to fasted rats varied from 0.56 to 1.7 μM/min [33, 65, 66]. In this chapter, we report the value of ca. 2.3

μM/min/g of liver in the interval 0-30 min, when glycogen synthesis prevailed over degradation.

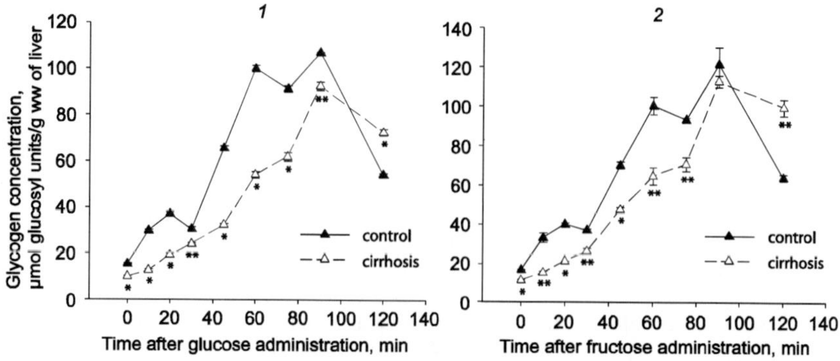

Figure 2. Dynamics of glycogen accumulation in normal and cirrhotic rat liver after refeeding with glucose (1) or fructose (2). *p<0.001, **P<0.01 – mean values were significantly different from control.

Contrary to glucose, fructose is mostly metabolised in the liver. Owing to fructokinase, a liver enzyme several times more active than glucokinase [67], fructose is easily absorbed by hepatic cells. It is thought that fructose, whose metabolism is not insulin-mediated, is used in the liver mostly for the synthesis of triglycerides and glycogen [68]. However, reports about the rate of glycogen synthesis from fructose are contradictory. In one study, the rate of glycogen accumulation in the human liver after fructose administration did not differ from that after glucose administration [69]. In another study, glycogen accumulated in the human liver several times faster after fructose infusion than after the infusion of glucose [70]. Comparative study of the glycogen accumulation dynamics in the liver of rats fasted for 24 h showed that 1 h after administration of glucose or fructose (4 g/kg of body weight) glycogen levels in the liver were the same. However, 2 h after refeeding with fructose glycogen content in the liver was ca. 1.2 times higher than after refeeding with glucose [34].

In our experiments, the dynamics and rates of glycogen accumulation in the rat liver after peroral administration of fructose did not differ from those after administration of glucose. Similar rates of glycogen accumulation after administration of glucose or fructose may be explained by several reasons, the main being that in rats most fructose is transformed into glucose already in the small intestine. This means that the researchers inevitably study the metabolism of glucose rather than fructose. The rates of this transformation vary in different animals and are much higher in rats than in humans.

Information about glycogen synthesis in the cirrhotic liver is scanty and ambiguous. On the one hand, the rate of glycogen synthesis from glucose in the liver of cirrhotic patients after nocturnal fasting did not differ from the control [10, 71]. On the other hand, glycogen content in the cirrhotic human and rat liver after a meal was lower than in the normal liver [1, 8, 21]. Our data indicate that glucose administration to fasted cirrhotic rats stimulated a rather fast glycogen accumulation in the liver. However, the rate of glycogen accumulation in the cirrhotic liver was, on the average, 26% lower than in the normal one. Substitution of glucose for fructose did not change the dynamics of glycogen accumulation in the cirrhotic rat liver.

A decreased rate glycogen accumulation in the cirrhotic liver may be due both to weakened glycogenesis and to enhanced glycogenolysis. A key role in these processes is played by glycogen synthase and glycogen phosphorylase, whose activity depends, first of all, on the degree of their phosphorylation and the concentration of allosteric modulators in the cells [72]. Peroral administration of glucose to control rats fasted for 48 h stimulates a fast increase in the GS activity [33, 64]. According to our data, 10 min after refeeding the activity of the enzyme increased by 95%. Fructose administration to fasted rats resulted in an even faster and more considerable increase in GS activity (194% after 10 min) than the administration of glucose (Figure 3). This agrees with the findings by Niewoehner et al. [34]. Some other authors, however, did not find any differences in the level of GS activity in the normal liver after refeeding of rats with glucose or fructose [73].

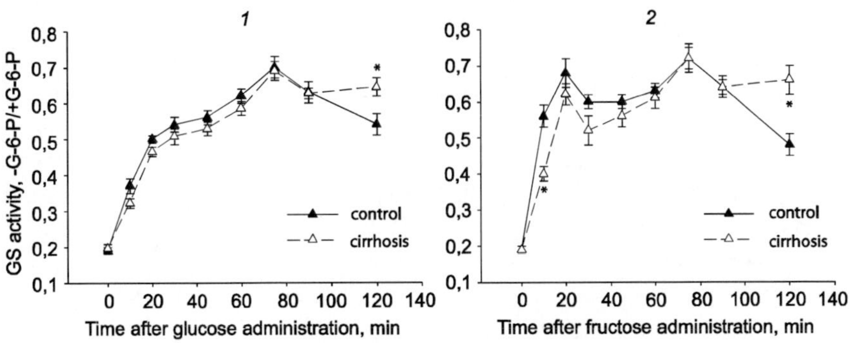

Figure 3. Activity of glycogen synthase (GS) in the normal and the cirrhotic rat liver at different time intervals after administration of glucose (1) or fructose (2) to fasted animals. *P<0.05 – mean values were significantly different from control.

The dynamics of GS activity in the cirrhotic liver after refeeding with glucose or fructose has not been studied. There are just a few contradictory reports about the activity of this enzyme at some (often uncertain) time after refeeding. For instance, the levels of total and active GS in the liver of fed rats with cirrhosis induced by bile duct ligation were almost twice as low as in the norm [56]. A considerable decrease in the GS activity in the liver, as compared to the control, 4 h after refeeding was observed in rats with cirrhosis induced by thioacetamide [21]. However, most evidence obtained on rats and humans indicates that in the post-absorptive period GS activity in the cirrhotic liver is the same as in the normal one [8, 74, 75].

Our research showed that administration of glucose stimulates a fast GS activation in the cirrhotic liver, however, not as fast as in the control. GS activity in the cirrhotic liver increased only by 63% 10 min after refeeding, reaching the control levels after 1.5-2 h. Administration of fructose to fasted cirrhotic rats resulted in a faster and more considerable increase in GS activity, up to the levels recorded in the control rats.

The main factor controlling hepatic glycogen metabolism is glycogen phosphorylase. It may be present in two forms, the active *a*-form (GP*a*) and the inactive *b*-form (GP*b*). GP*a* is a strong inhibitor of phosphatase synthase, the decrease in its activity triggering GS activation and glycogen synthesis [76]. Therefore, it has previously been thought that GP*a* inactivation by transformation into inactive dephosphorylated *b*-form is a necessary condition for the start of glycogen synthesis in the liver [77].

Reciprocal interaction between GS and GP during glycogen synthesis has been often shown [29, 78]. At the same time, it was shown that during glycogen synthesis GS and GP may be highly active simultaneously [33, 34, 64, 79]. Some authors came to the conclusion that reciprocal interaction between GS and GP is expressed only after refeeding of fasted rats with glucose, while after refeeding with fructose both enzymes are equally and highly active [80]. High level of activity of both GS and GP during glycogen synthesis and the lack of correlation between GS activity and the rate glycogen accumulation in the liver indicate that many aspects of glycogen synthesis are far from being clear.

In our research, GP*a* activity in the normal liver after peroral administration of glucose to rats fell fast, after 20 min making up ca. 50% of the GP*a* activity in the liver of fasted animals (Figure 4).

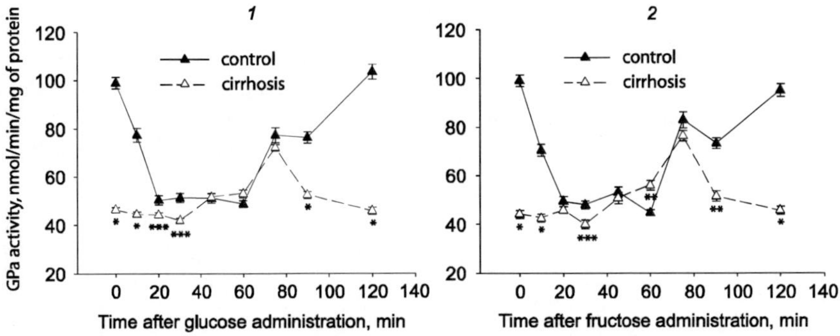

Figure 4. Activity of glycogen phosphorylase a (GPa) in the normal and the cirhhotic rat liver at different time intervals after administration of glucose (1) or fructose (2) to fasted animals. *P<0.001, **P<0.01, ***P<0.05 – mean values were significantly different from control.

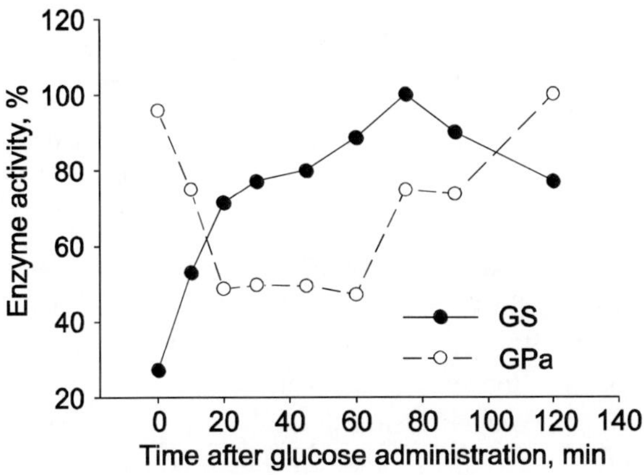

Note: Maximum enzyme activity is taken as 100%.

Figure 5. Activity of glycogen phosphorylase a (GPa) and glycogen synthase (GS) in the normal rat liver at different time intervals after administration of glucose to fasted animals.

GS and GP activity dynamics was reciprocal in the interval 0-30 min. Later, however, though GS activity remained high, that of GP*a* increased up to the initial level, indicating that later still (after 60 min) glycogen was accumulated in the liver with both enzymes being highly active (Figure 5). As a consequence, the activity of these enzymes and glycogen content in the liver was strongly correlated at the early stages of refeeding with glucose (0-30 min) and not correlated at all later on (45-120 min) (Figure 6). The dynamics of the GP*a* activity in the normal liver after administration of fructose to fasted rats was the same as after that of glucose. Thus, our data disagree with the data of [33, 34], who reported that the GP*a* activity did not decrease after administration of glucose or fructose to rats fasted for 24 h, the concentrations of monosaccharides being similar to those used in our research.

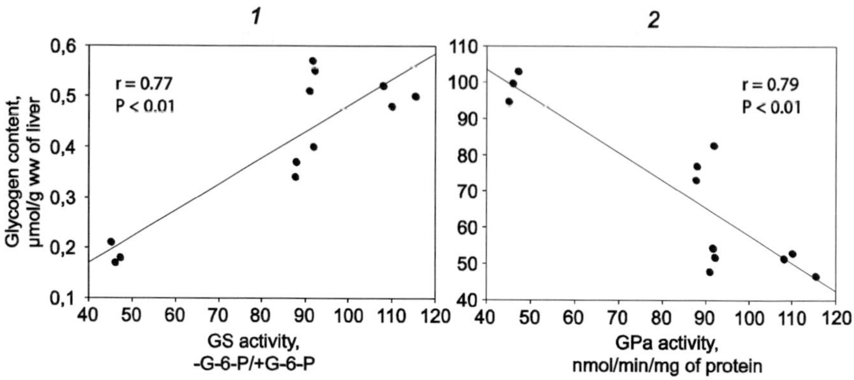

Figure 6. Dependence between glycogen content in hepatic cells and activity of glycogen synthase (1) and glycogen phosphorylase a (2) in the normal rat liver during the first 30 min after refeeding.

Contrary to these data, in our research GP*a* activity in the liver of fasted rats treated with CCl_4 for 6 months was low. After peroral administration of glucose or fructose, GP*a* activity, except for a small increase at 75 min after refeeding, remained at the same low level as in the liver of fasted rats. As a result, initiation of glycogen synthesis and its further accumulation in the cirrhotic liver proceeded at the background of a constantly low GP*a* activity (ca. 50% of the values in the normal liver of fasted rats). Noteworthy, we also observed the decreasing level of total GP during cirrhosis. This seems to indicate a weakened synthesis of this enzyme in the cirrhotic liver.

High GS and GP activities during transition from fasted to fed state, when glycogen is being accumulated in hepatic cells, indicate that glycogen is both synthesized and degraded at that time. One may suppose that the increase in the glycogen concentration in the liver is an important factor stimulating glycogen turnover. Otherwise, glycogen accumulation in hepatic cells would have been uncontrolled. Dependence between glycogen concentration and glycogenolysis rate in the liver also indicates the presence of glycogen turnover [79]. Besides, there is some direct evidence of the dependence between the rate of glycogen turnover and its content in the liver [79, 81].

GS activity has often been shown to depend on the concentration of glucose-6-phosphate in the cells. It is mostly formed during gluconeogenesis or as a result of glucose phosphorylation with the participation of glucokinase [82]. GK was shown to exercise considerable control over glycogen synthesis from glucose in hepatic cells [83].

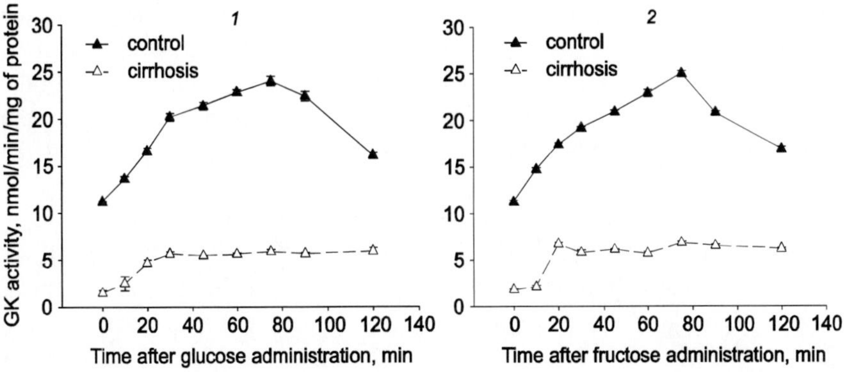

Figure 7. Activity of glucokinase in the normal and the cirrhotic rat liver ar different time intervals after administration of glucose (1) or fructose (2). All mean values were significantly different from control – P<0.001.

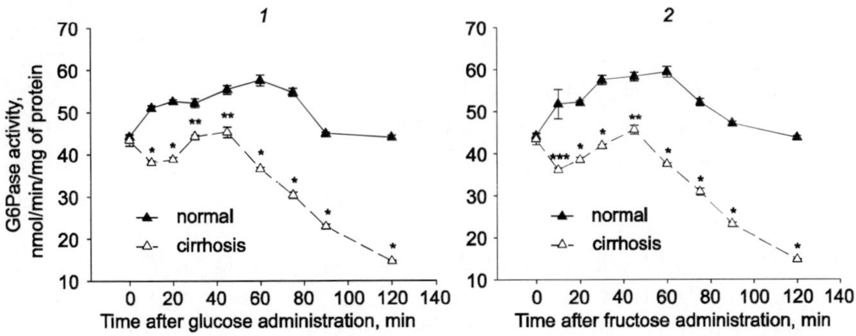

Figure 8. Activity of glucose-6-phosphatase (G6Pase) in the normal and the cirrhotic rat liver at different time intervals after administration of glucose (1) or fructose (2) to fasted animals. *P<0.001, **P<0,01, ***P<0.05 – mean values were significantly different from control.

Peroral administration of glucose to control rats resulted in a 20% increase in GK activity after 10 min. GK levels almost doubled 60 min after refeeding with glucose but started to fall after 75 min. A similar dynamics of GK activity was observed after administration of fructose to fasted rats. This observation lend support to the idea that most of the perorally administered fructose is transformed into glucose in the intestine (Figure 7).

GK activity was much lower in the cirrhotic liver as compared to the norm. In the liver of fasted cirrhotic rats it made up only 16-17% of the level observed in the liver of fasted control animals. This is in agreement with the data of other authors [7, 8, 27]. Though GK levels increase after peroral administration of glucose or fructose to fasted rats, its maximum activity in the cirrhotic liver remains for a long time at a level several times lower than the normal one. These results indicate a considerable and stable decrease of the glucose phosphorylation rate in the cirrhotic liver.

GK and G6Pase play a crucial role in the absorption and release of glucose by the liver. In our research, G6Pase activity during refeeding of normal rats with glucose or fructose grew slowly for 60 min, but after reaching the peak (30-35% higher than G6Pase activity in the liver of fasted rats) reverted to the initial level (Figure 8).

Our results indicate that gluconeogenesis and glucose production continue during 120 min of refeeding. We found that increasing insulin concentrations after peroral administration of glucose to rats fasted for 24 h did not result in a decrease in the glucose levels in the liver. This result indicates that gluconeogenesis continues at least during 2 h after administration of glucose and that glucose production proceeds without interruptions despite the absorption of exogenous glucose by the liver and the use of gluconeogenesis substrates for glycogen synthesis [84, 85].

On the contrary, Minassian et al. (1999) showed that refeeding of rats with rough food resulted in a progressive inhibition of G6Pase activity [29]. It was suggested that G6Pase inhibition may be necessary for glycogen accumulation in the liver [66, 86]. It should be noted, however, that, firstly, the conclusion about inhibition of G6Pase activity during refeeding was based on the analysis of the dependence between G6Pase activity and glycogen content in the liver constructed with the use of only three time points: 90, 180 and 360 min after refeeding. Secondly, it seems improbable that the number of gluconeogenesis substrates, formed mostly in the muscles, decreased after a meal. On the contrary, it is much more likely that the amount of lactate, pyruvate and alanine transported to the liver during refeeding increase. This is associated with the active reparative processes which attempt restor the normal functional activity of various tissues undermined by fasting.

Contrary to the normal liver, where G6Pase activity increased after the administration of monosaccharides, in the cirrhotic liver it started to fall abruptly 45 min after refeeding. As a result, 120 min after administration of glucose or fructose G6Pase activity was ca. 3 times lower than in fasted rats (Figure 8).

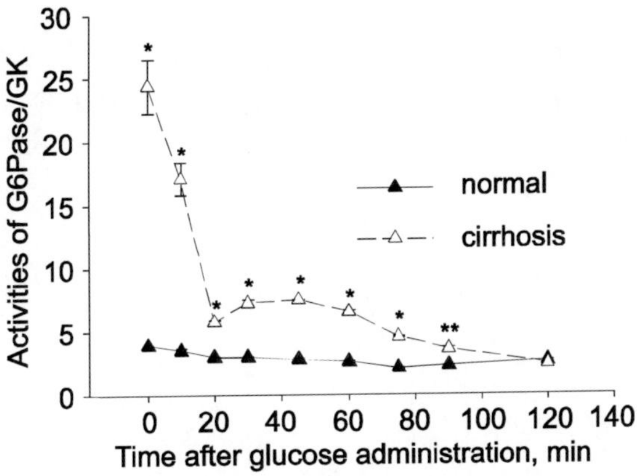

Figure 9. Proportion of activities of glucose-6-phosphatase (G6Pase) and glucokinase (GK) in the normal and the cirrhotic rat liver at different time intervals after administration of glucose. *P<0.001, **P<0.01 - mean values were significantly different from control.

Under certain conditions glucose is known to be phosphorylated in the liver not with the help of glucokinase or hexokinase but with the help of G6Pase. The latter is sometimes referred to as the third glucokinase. In case of experimental diabetes or after prolonged fasting, when GK activity was very low, glucose was actively phosphorylated by transferase activity of G6Pase [87]. Therefore, it may be supposed that in the cirrhotic liver, when much glucose comes from the intestine and GK activity is dramatically decreased, G6Pase steps forward to perform glucose phosphorylation. A shift from hydrolase to transferase function of G6Pase may be considered as an adaptive response to a weak phosphorylating activity of GK during cirrhosis.

The ratio of G6Pase and GK activities is thought to reflect the level of glucose absorption in the liver [88], making it possible to assess the ratio of glucose absorption and production. Our data indicate that the substrate flow through G6Pase is much higher that that through GK both in the control and in the experimental rats (Figure 9). In the normal liver, the substrate flow during refeeding, despite its relative decrease as compared to fasted animals, is most channelled to glucose production. In the cirrhotic liver, it is also so at the early stages of refeeding (0-45 min) but ca. 1 h after refeeding, as the glucose-phosphorylating capacity of the liver increases (hydrolase/transferase shift in G6Pase activity + GK activity 4 times higher than in fasted rats), glucose absorption in the liver starts to prevail over its release.

Our results indicate that in the normal and the cirrhotic liver during refeeding, or at least during 2 h after its start, the activity of glycolysis and gluconeogenesis enzymes (GK and PFK, Fru-1,6-Pase and G6Pase) is simultaneously maintained at a high level. At the same time, it should be noted that in the normal liver changes in the activity of glycolysis and gluconeogenesis enzymes during refeeding are well correlated (Figure 10).

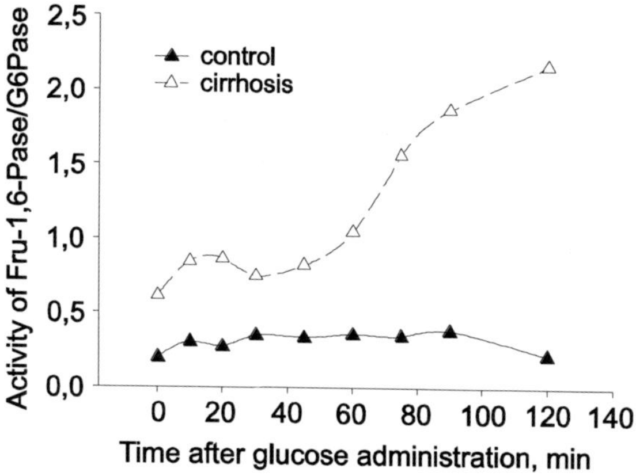

Figure 10. Proportion of activities of fructose-1,6-diphosphatase (Fru-1,6-Pase) and glucose-6-phosphatase (G6Pase) in the normal and the cirrhotic rat liver at different time intervals after administration of glucose.

After a meal, 25-40% of glucose is absorbed by the liver, the rest being mostly used by muscles, brain, erythrocytes and adipose tissue [2, 89]. Liver-absorbed glucose may be used in a non-oxidative way for glycogen synthesis, oxidated during glycolysis with the formation of lactate and CO_2 or be used for fatty acids synthesis.

The relative contribution of direct (from glucose) and indirect (through gluconeogenesis) pathways into glycogen accumulation in hepatic cells is a subject of debate. Some authors think that the indirect pathway contributed more to glycogen synthesis [2, 17], while the others hold an opposite viewpoint [90]. According to our results, glycogen synthesis in the normal liver after peroral administration of glucose or fructose to rats during the first 60 min is direct. Later, however, the relative importance of indirect pathway grows. It seems to prevail in the post-absorptive period, becoming the only available option during fasting, when glycogen reserves are depleted. In the cirrhotic liver, owing to the very low GK activity and a high gluconeogenesis level, the main role in glycogen accumulation during fasting and at all stages after refeeding seems to be played by the indirect pathway.

To sum up, cirrhosis is accompanied by profound metabolic disorders of glycogen in the liver. They are expressed, first of all, in decreased rates of glycogenesis and glycogenolysis in hepatic cells. This is due to a lower activity of glycogen phosphorylase, glucokinase and G6Pase in the cirrhotic liver. Disorders of glycogen metabolism in the cirrhotic liver are triggered by reconstructions of parenchyma and the liver bloodstream, leading to an enhancement of lipid peroxidation in the membrane complexes of hepatic cells, and to hypoxia.

POST-ABSORPTIVE PERIOD

Though experimental liver cirrhosis is strikingly similar in rats and humans [91, 92, 93] data obtained on animals can be used in medical practice only with the greatest caution [94]. The risk of extrapolating animal data to humans is associated not only with the differences in biology, metabolism or other species-specific characters but also with the fact that, contrary to experimental animals, there are no ways of obtaining detailed information about the metabolism in a particular human organ. Therefore, the researchers usually have to be content with comparison of detailed information about metabolism obtained on experimental animals with the relatively scanty information from human metabolic research.

Noteworthy, most available information about carbohydrate metabolism in cirrhotic liver of humans was obtained after nocturnal fasting, that is, relates to post-absorptive period. In order to extrapolate with greater certainly the data on hepatic glycogen metabolism of rats to humans, we conducted a comparative study of glycogen metabolism in the human and rat liver in the post-absorptive period of the food cycle (8-12 h after the end of meal). In this period degradation of glycogen prevails over its synthesis in the liver [64] while glucose production proceeds both by gluconeogenesis and glycogenolysis.

In our cytophotometric research, glycogen content in the cirrhotic liver of rats and humans in the post-absorptive period was, respectively, 3.2 and 2.6 times higher than in the normal one (Figure 11).

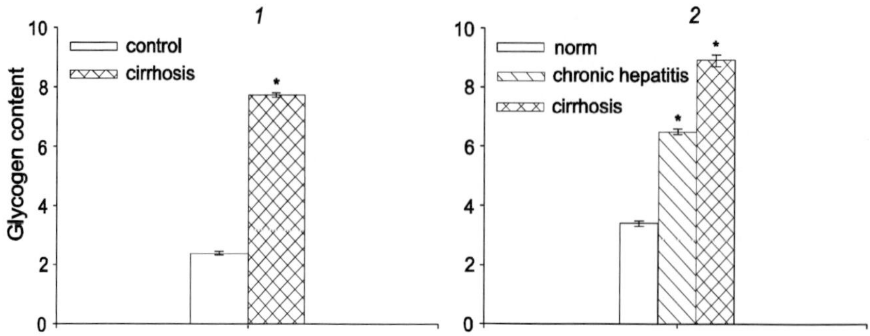

Figure 11. Hepatic glycogen content (arbitrary units in control and experimental rats (1); patients without liver disease, patients with chronic hepatitis and liver cirrhosis (2). *P<0.001 - mean values were significantly different from control.

Besides, in agreement with the earlier data [11], glycogen content in hepatic cells of patients with chronic hepatitis increased with the increasing impairment of the liver.

A considerable increase in glycogen content in the cirrhotic human and rat liver in the post-absorptive period may reflect both the decreased capacity of hepatic cells to glycogen degradation and an increased rate of glycogen synthesis in the cells. GS activity in the liver of rats with CCl_4-induced cirrhosis in the post-absorptive period was slightly higher (by 7.5 %) than in normal animals.

On the other hand, GS activity in cirrhotic human liver after nocturnal fasting did not differ from the norm, which agrees with the data of Kudryavtseva and co-authors [11, 95]. Moreover, Krahenbuhl and co-authors showed that GS activity in the liver of rats with biliary liver cirrhosis induced by bile duct ligation was lower than in the normal liver. A considerable fall in the total and active GS forms in this case correlated with the decrease in the number of hepatic cells [56]. However, in the case of CCl_4-induced liver cirrhosis in rats, the same authors demonstrated that though the total GS in the liver increased, the levels of its active form did not differ from the norm [8]. Taken into account these data and the results of the research reported in this chapter on the lower rate of glycogen accumulation in the cirrhotic rat liver after refeeding with glucose or fructose as compared to the norm, we may conclude that the rate of glycogen synthesis in the cirrhotic liver is, at least, not higher than in the normal one.

Little is known about the rate of glycogen degradation in the cirrhotic liver, the existing reports being few and contradictory. On the one hand, glycogen degradation was reported to be faster during cirrhosis than in the norm [8], but on the other hand, the rate of glycogenolysis in the liver of cirrhotic patients was shown to be by 66% lower than in the normal liver [7].

GP is the key enzyme of glycogen degradation. However, information about its activity in the cirrhotic liver is scattered and uncertain. Total and active GP levels in the liver of CCl_4-cirrhotic rats and in the normal liver were the same [8], but in the liver of rats with biliary cirrhosis the level of both GP forms was decreased [56]. GP activity was decreased by 10% in the liver of 3 (out of 35) children with chronic liver diseases [96].

Our research indicates that GP*a* activity in the cirrhotic liver of rats in the post-absorptive period by 12.4 % (Figure 12). This agrees with our earlier results that GP*a* activity in the cirrhotic rat liver decreases by 15-35% as compared to the norm [74, 97].

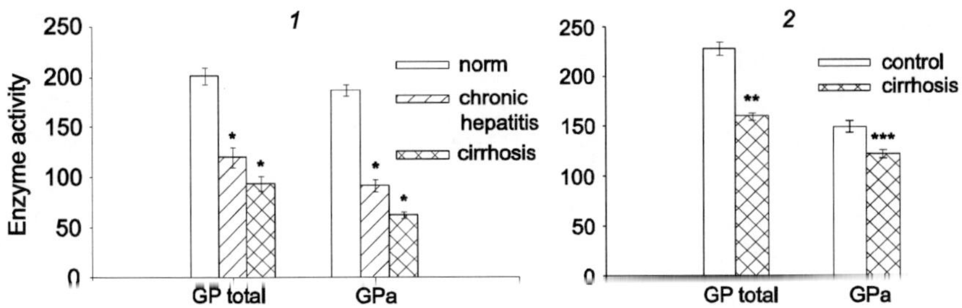

Figure 12. Activity of total glycogen phosphorylase (GP total), a-form of glycogen phosphorylase (GPa) (nmol/min/mg of protein) in the liver of patients without liver disease, patients with chronic hepatitis and liver cirrhosis (1); In the normal and the cirrhotic rat liver (2). *P<0.001, **P<0.01, ***P<0.05 – mean values were significantly different from control.

GP*a* activity in the cirrhotic human liver after nocturnal fasting was lower than in the cirrhotic liver of rats. Interestingly, it depends on the stage of the chronic hepatitis (Figure 12), making up 49% of the norm at the pre-cirrhosis stage and only 33% at the cirrhotic stage. This agrees well with the results obtained earlier in our laboratory [98]. Noteworthy, a decrease in the GP*a* activity was accompanied by a considerable decrease in the total GP activity, which made up 29% in cirrhotic rat liver and 53% in cirrhotic human liver. In support of these data, it was also found that mRNA expression of glycogen phosphorylase in the cirrhotic human liver decreased by 50% [47].

It is unclear why GP activity should decrease in the cirrhotic liver. An important role may be played by the concentration of various co-factors and modulators of GP activity. One such modulator is cAMP, whose concentration in the liver is determined by the activity of adenylate cyclase. Its stimulation results in the increase in the cAMP concentration in the cell, leading to an increase in the activity of protein kinase, under whose impact the inactive GP*b* is transformed into GP*a*. Adenylate cyclase activity and cAMP concentration in the cirrhotic liver were shown to be considerably lower than the norm [59; 99], which may bring about the decrease in GP activity.

GP activity is also influenced by the concentration of vitamin B_6. Pyridoxal phosphate (B_6) is part of GP molecules [57]. In muscles, 50-86% of B_6 is bound with glycogen phosphorylase, thus, the pool of this enzyme contains much of the organism's B_6. A study of GP denaturation kinetics showed that this enzyme is always inactive in the absence of B_6. Only when the vitamin is present, the enzyme acquires the conformation necessary for the manifestation of the maximum activity [100]. GP activity was shown to be considerably decreased during B_6-avitaminosis. GS activity during B_6-avitaminosis is unchanged [100].

Pyridoxal phosphate content in the liver of cirrhotic patients was shown to be 2-3 times lower than in the normal liver [101]. Therefore, it is possible that the decreasing content of pyridoxal phosphate in the cirrhotic liver may play the key role in the decrease of GP*a* activity.

Regardless of the mechanisms of the decrease in GP activity in the cirrhotic liver, most available data indicate that the rate of glycogen degradation in the cirrhotic liver is lower than in the normal one. This may be one of main reasons why glycogen content in the cirrhotic liver in the post-absorptive period is increased.

Inhibition of G6Pase activity in the liver of fasted rats was shown to channel the glucose-6-phosphate flow in hepatic cells towards glycogen synthesis rather than glucose production [29, 102]. Based on this, we may suppose that the change in the G6Pase activity would considerably influence post-absorptive glycogen levels in the liver. Most data on G6Pase activity were obtained on the liver of humans after nocturnal fasting or the liver of rats in the post-absorptive period. G6Pase activity in the liver of CCl_4-cirrhotic rats was sharply decreased, making up only 23-36% of the G6Pase activity in the normal rat liver [75, 97, 98]. In our research, G6Pase activity in the cirrhotic rat liver in the post-absorptive period made up only 26% of its activity in the normal liver (Figure 13). A considerable decrease in G6Pase activity was also recorded in the cirrhotic human liver [6, 17, 46, 103]. In a study using biopsied human liver material it was shown that G6Pase activity depended on the degree of liver affection: during pre-cirrhosis chronic hepatitis G6Pase activity was decreased by 1.2 times as compared to the norm while at the cirrhotic stage, by 4.4 times [11]. Similar values of G6Pase activity in the liver of patients with chronic hepatitis and liver cirrhosis were obtained in our research (Figure 13).

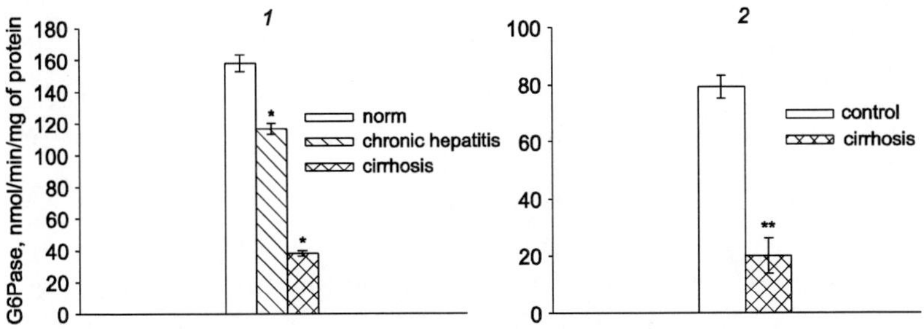

Figure 13. Activity of glucose-6-phosphatase (G6Pase) in the liver of patients without liver disease, patients with chronic hepatitis and liver cirrhosis (1); In the liver of normal and cirrhotic rats (2); $^*P<0.001$, $^{**}P<0.005$ – mean values were significantly different from control.

Thus, the results of biochemical and cytophotometric analysis showed that the decreased G6Pase activity, alongside with the decrease in the GP*a* activity, may be the major reason of glycogenosis in the cirrhotic liver of rats and humans in the post-absorptive period.

The reasons of the decreased G6Pase activity in the cirrhotic liver are unclear. Such a sharp fall cannot be explained only by the decreasing proportion of parenchyma in the cirrhotic liver. Many authors think that one of the main reasons of the G6Pase system failure is the damage of RER membranes by the products of lipid peroxidation [104, 105, 106]. In this connection, the decreased activity of superoxide dismutase and catalase and, thus, the increase of the level of lipid peroxidation products in the blood and liver of cirrhotic rats and humans may indicate that the enhancement of lipid peroxidation is one of the main mechanisms responsible for the fall of G6Pase activity in the cirrhotic liver. Whatever the case, low G6Pase activity for at least 12 h indicates decreased glucose production by the cirrhotic liver.

CONCLUSION

To sum up, hepatic carbohydrate metabolism in cirrhotic rats and humans is dramatically different from the normal one. The main differences are an abrupt fall in the activities of GP*a* and G6Pase, which continue to decrease as the liver is further impaired, and a fall in the activity of GK, accompanied by a rise in the activity of Fru-1,6-Pase and PFK. The results presented in this chapter indicate that gluconeogenesis is the key process in the liver during starvation, proceeding on the background of low glycogen reserves in hepatic cells and low blood glucose. Its rate is increased in the cirrhotic liver as compared to the normal one.

REFERENCES

[1] Riggio, O., Merli, M., Leonetti, F., Giovannetti, P., Foniciello, M., Folino, S., & Capocaccia, L. (1997). Impaired nonoxidative glucose metabolism in patients with liver cirrhosis: effects of two insilin doses. *Metabolism*, *46*, 840-843.

[2] Kruszynska, Y. T. (1999). Functions of the liver: Carbohydrate metabolism. In J. Bircher, J.-P. Benhamou, N. McIntyre, M. Rizzetto, & J. Rodes (Eds.), *Oxford textbook of clinical hepatology* (pp.257-286). Oxford, UK: Oxford University Press.

[3] Radziuk, J., & Inculet, R. (1983). The effects of ingested and intravenous glucose on forearm uptake of glucose and glucogenic substrate in normal man. *Diabetes*, *32*, 977-981.

[4] Kruszynska, Y. T., Harry, D. S., Fryer, L. G., & McIntyre, N. (1994). Lipid metabolism and substrate oxidation during intravenous fructose administration in cirrhosis. *Metabolism*, *43*, 1171-1181.

[5] Felig, P., Wahren, J., & Hendler, R. (1975). Influence of oral glucose ingestion on splanchnic glucose and gluconeogenic substrate metabolism in man. *Diabetes*, *24*, 468-475.

[6] Owen, O. E., Reichle, F. A., Mazzoli, M. A., & Kreulen, I. (1981). Hepatic, gut and renal substrate flux in patients with hepatic cirrhosis. *J. Clin. Invest.*, *68*, 240-252.

[7] Petersen, K. F., Krssak, M., Navarro, V., Chandramouli, V., Hundal, R., Schumann, W. C., & Shulman, G. I. (1999). Contributions of net hepatic

glycogenolysis and gluconeogenesis to glucose production in cirrhosis. *Am. J. Physiol.*, *276*, E529-E535.

[8] Krahenbuhl, S., Weber, F. L., & Brass, E. P. (1991). Decreased hepatic glycogen content and accelerated response to starvation in rats with carbon tetrachloride-induced cirrhosis. *Hepatology*, *14*, 1189-1195.

[9] Bondy, P. K. (1958). Abnormalities in liver disease. *Am. J. Med.*, *24*, 428-436.

[10] Schneiter, P., Gillet, M., Chiolero, R., Jequier, E., & Tappy, L. (1999). Hepatic nonoxidative disposal of an oral glucose meal in patients with liver cirrhosis. *Metabolism*, *48*, 1260-1266.

[11] Kudryavtseva, M. V., Sakuta, G. A., Scorina, A. D., Stein, G. I., Emelyanov, A. V., & Kudryavtsev, B. N. (1996). Quantitative analysis of glycogen content in hepatocytes of portal and central lobule zones of normal human liver and in patients with chronic hepatitis of different etiology. *Tissue Cell*, *28*, 279-285.

[12] Newgard, C. B., Foster, D. W., & McGarry, J. D. (1984). Evidence for suppression of hepatic glucose-6-phosphatase with carbohydrate feeding. *Diabetes*, *33*, 192-195.

[13] Sugden, M. C., Watts, D. I., Palmer, T. N., & Myles, D. D. (1983). Direction of carbon flux in starvation and after refeeding in vitro and in vivo effects of 3-mercaptopropionate. *Biochem. Int.*, *7*, 329-337.

[14] Piniewska, D. M., McCulloch, A. J., Bramble, M. G., Taylor, R., Record, C. O., & Alberti, K. G. (1986). Glucose turnover in compensated hepatic cirrhosis. *Horm. Met.*, *18*, 834-837.

[15] Owen, O. E., Trapp, V. E., Reichard, G. A. Jr., Mozzoli, M. A, Moctezuma, J., Paul, P., & Boden, G. (1983). Nature and quantity of fuels consumed in patients with alcoholic cirrhosis. *J. Clin. Invest.*, *72*, 1821-32.

[16] Krahenbuhl, S., & Reichen, J. (1993). Decreased hepatic glucose production in rats with carbon tetrachloride-induced cirrhosis. *J. Hepatol*, *19*, 64-70.

[17] Kruszynska, Y. T., & McIntyre, N. (1991). Carbohydrate metabolism. In: *N. McIntyre*, P. J. Benhamou, J. Bircher, M. Rizzetto & J. Rodes (Eds.), *Oxford textbook of clinical hepatology (pp.129-143). Oxford, UK: Oxford University Press.*

[18] Bois-Joyeux, B., Chanez, M., & Peret, J. (1990). Age-dependent glycolysis and gluconeogenesis enzyme activities in starved-refeed rats. *Diabete & Metabolism (Paris)*, *16*, 504-512.

[19] Jungermann, K. (1992). Role of intralobular compartmentation in hepatic metabolism. *Diabete & Metabolisme*, *18*, 81-86.

[20] Avdeeva, L. V., & Vorobyova S. A. (2005). Hormonal regulation of metabolism and functions of the organism. In E. S. Severin (Eds.), *Biochemistry* (pp. 545-616). Moskow: GEOTAR Media.

[21] Giardina, M. G., Matarazzo, M., & Sacca, L. (1994). Kinetic analysis of glycogen sunthase and PDC in cirrhotic rat liver and skeletal muscle. *Am. J. Physiol.*, *267*, E900-E906.

[22] Mlekusch, W., Paletta, B., Truppe, W., Paschke, E., & Grimus, R. (1981). Plasma concentration of glucose, corticosterone, glucago and insulin and liver content of metabolic substrates and enzymes during starvation and additional hypoxia in tha rat. *Horm. Metab. Res.*, *13*, 612-614.

[23] Calder, P. C., & Geddes, R. (1992). Heterogeneity of glycogen synthesis upon refeeding following starvation. *Int. J. Biochem*, *24*, 71-77.

[24] Johansson, U., Wahren, J., & Eriksson, L. S. (1994). Splanchnic and peripheral glucose metabolism in cirrhosis. *J. Hepatol.*, *20*, 760-767.

[25] Brown, K. S., Kalinowski, S. S., Megill, J. R., Durham, S. K., & Mookhtiar, K. A. (1997). Glucokinase regulatory protein may interact with glucokinase in the hepatocyte nucleus. *Diabetes*, *46*, 179–186.

[26] Brocklehurst, K. J., Payne, V. A., Davies, R. A., Carroll, D., Vertigan, H. L., Wightman, H. J., & Agius, L. (2004). Stimulation of hepatocyte glucose metabolism by novel small molecule glucokinase activators. *Diabetes*, *53*, 535-541.

[27] Lowes, W., Walker, M., Alberty, K. G., & Agius L. (1998). Hexokinase isoenzymes in normal and cirrhotic human liver: suppression of glucokinase in cirrhosis. *Biochem. Biophys. Acta*, *1379*, 134-142.

[28] Bois-Joyeux, B., Chanez, M., & Peret, J. (1990). Age-dependent glycolysis and gluconeogenesis enzyme activities in starved-refed rats. *Diabete Metab.*, *16*, 504-512.

[29] Minassian, C., Montano, S., & Mithieux, G. (1999). Regulatory role of glucose-6-phosphatase in the repletion of the liver glycogen during refeeding in fasted rats. *Biochem. Biophys. Acta*, *1452*, 172-178.

[30] Claus, T. H., Nyfeler, F., Muenkel, H. A., Burns, M. G., & Pilkis, S. J. (1984). Changes in fructose-2,6-bisphosphate levels after glucose loading of starved rats. *Biochem. Biophys. Res. Commun.*, *122*, 529-534.

[31] Ercan, N., Gannon, M. C., & Nuttall, F. Q. (1994). Liver glycogen synthase, phosphorylase, and the glycogen concentration in rats given a

glucose load orally: a 24-hour study. *Arch. Biochem. Biophys.*, *315*, 35-40.

[32] Roach, P. J. (2002). Glycogen and its metabolism. *Curr. Mol. Med.*, *2*, 101-120.

[33] Katz, J., Golden, S., & Wals, P. A. (1979). Glycogen synthesis by rat hepatocytes. *Biochem. J.*, *180*, 389-402.

[34] Niewoehner, C. B., Gilboe, D. P., & Nuttall F. Q. (1984). Metabolic effects of oral glucose in the liver of fasted rats. *Am. J. Physiol.*, *246*, E89-E94.

[35] Niewoehner, C. B., Gilboe, D. P., Nuttall, G. A., & Nuttall, F. Q. (1984). Metabolic effects of oral fructose in the liver of fasted rats. *Amer. J. Physiol.*, *247*, E505-E512.

[36] Madar, Z. (1989). Pyruvate dehydrogenase and glycogen synthase activity at transition from fasted to feed state. *Biochem. Med. Metab. Biol.*, *41*, 93-104.

[37] Biorn, A. C., & Graves, D. J. (2001). The amino-terminal tail of glycogen phosphorylase is a switch for controlling phosphorylase conformation, activation, and response to ligands. *Biochemistry*, *40*, 5181-5189.

[38] Roach, P. J., Depaoli-Roach, A. A., Hurley, T. D., & Tagliabracci, V.S. (2012). Glycogen and its metabolism: some new developments and old themes. *Biochem. J.*, *441*, 763-787.

[39] Exton, J. H., Ui, M., & Park, C. R. (1970). Mechanism of glucagon action on gluconeogenesis. *Hoppe Seylers Z. Physiol. Chem.*, *351*, 289-290.

[40] Merli, M., Eriksson, L. S., Hagenfeldt, L., & Wahren, J. (1986). Splanchnic and leg exchange of free fatty acids in patients with liver cirrhosis. *J. Hepatol.*, *3*, 348-355.

[41] Shmueli, E., Walker, M., Alberti, G., & Record, C. O. (1993). Normal splanchnic but impaired peripheral insulin-stimulated glucose uptake in cirrhosis. *Hepatology*, *18*, 86-95.

[42] Petrides, A. S., Vogt, Ch., Schulze-Berge, D., Matthews, D., & Strohmeyer, G. (1994). Pathogenesis of glucose intolerance and diabetes mellitus in cirrhosis. *Hepatology*, *19*, 616-627.

[43] Waddel, I. D., & Burchell, A. (1991). Transverse topology of glucose-6-phosphatase in rat hepatic endoplasmic reticulum. *Biochem. J.*, *275*, 133-137.

[44] James, J., Bosch, K. S., Zuyderhoudt, F. M., Houtkooper, J. M., & van Gool, J. (1985). Histophotometric estimation of volume density of

collage as an indication of fibrosis in rat liver. *Histochemistry*, *85*, 129-133.

[45] Reichen, J., Arts, F., Schafroth, U., Zimmermann, A., Heltner, Th. B., & Zysset, T. (1987). Aminopyrine N-Demethilation by rats with liver cirrhosis. *Gastroenterology*, *93*, 719-726.

[46] Gressner, A. M., & Schuppan, D. (1999). Cellular and molecular pathobiology, pharmacological intervention, and biochemical assessment of liver fibrosis. In J. Bircher, J.-P. Benhamou, N. McIntyre, M. Rizzetto, & J. Rodes (Eds.), *Oxford textbook of clinical hepatology* (pp.607-627). Oxford, UK: Oxford University Press.

[47] Sotaniemi, E. A., Saami, H. U., Arranto, A. J., & Rautio, A. (1985). Carbohydrate intolerance associated with reduced hepatic glucose phosphorylating and releasing enzyme activities and peripheral insulin resistence in alcoholics with liver cirrhosis. *J. Hepatol.*, *1*, 277-287.

[48] Krahenbuhl, L., Lang, C., Ludes, S., Seiler, C., Shafer, M., Zimmermann, A., & Krahenbuhl, S. (2003). Reduced hepatic glycogen stores in patients with liver cirrhosis. *Liver Int.*, *23*, 101-109.

[49] Mion, F., Galoun, A., Agosto, E., & Minaire, Y. (1996). Carbon tetrachloride-induced cirrhosis in rats: influence of the acute effects of the toxin on glucose metabolism. *Hepatology*, *23*, 582-588.

[50] Aiston, S., Hampson, L., Gomez-Foix, A. M., Guinovart, J. J., & Agius, L. (2001). Hepatic glycogen synthesis is highly sensitive to phosphorylase activity: evidence from metabolic control analysis. *J. Biol. Chem.*, *276*, 23856 –23866.

[51] Wu, C., Okar, D. A., Newgard, C. B., & Lange, A. J. (2002). Increasing fructose 2,6-bisphosphate overcomes hepatic insulin resistance of type 2 diabetes. *Am. J. Physiol.*, *282*, E38-E45.

[52] Wu, C., Okar, D. A., Kang, J., & Lange, A. J. (2005). Reduction of hepatic glucose production as a therapeutic target in the treatment of diabetes. *Current Drug Targets: Immune, Endocrine & Metabolic Disorders*, *5*, 51-59.

[53] Rossetti, L., Giaccari, A., Barzilai, N., Howard, K., Sebel, G., & Hu, M. (1993). Mechanism by which hyperglycemia inhibits hepatic glucose production in conscious rats. Implications for the pathophysiology of fasting hyperglycemia in diabetes. *J Clin Invest.*, *92*, 1126-1134.

[54] Iynedjian, P. B., Marie, S., Gjinovci, A., Genin, B., Deng, S. P., Buhler, L., & Mentha, G. (1995). Glucokinase and cytosolic phosphoenolpyruvate carboxykinase (GTP) in the human liver.

Regulation of gene expression in cultured hepatocytes. *J. Clin. Invest.*, *95*, 1966-1973.

[55] Magnuson, M. A. (1990). Glucokinase gene structure. Functional implications of molecular genetic studies. *Diabetes*, *39*, 523–527.

[56] Agius, L. (1994). Hexokinase and glucokinase binding in permeabilized guinea-pig hepatocytes. *Biochem J.*, *303*, 841-846.

[57] Krahenbuhl, L., Talos, C., Reichen, J., & Krahenbuhl, S. (1996). Progressive decrease in tissue glycogen content in rats with long-term cholestasis. *Hepatology*, *24*, 902-907.

[58] Rath, V. L., Ammirati, M., LeMotte, P. K., Fennel, K. F., Mansour, M. N., Danley, D. E., & Pandit, J. (2000). Activation of human liver glycogen phosphorylase by alteration of the secondary structure and packing of the catalytic core. *Mol. Cell.*, *6*, 139-148.

[59] Francavilla, A., Jones, A. F., & Starzl, T. E. (1978). Cyclic AMP metabolism and adenylate cyclase concentration in patients with advanced hepatic cirrhosis. *Gastroenterology*, *75*, 1026-1032.

[60] Vysotskaya, R. A., Loginov, A. S., Varvanina, G. G., & Pilenitsyn, A. Yu. (1998). The value of adenylate cyclase system of the liver in the development of its chronic lesions. *Byul. Eksp. Biol. Med.*, *125*, 450-453.

[61] Stratton, R. J., Green, C. J., & Elia, M. (2003). *Disease-Related Malnutrition: an Evidence-Based approach to treatment*. Wallingford: CABI Publishing.

[62] Castilla-Cortazar, I., Prieto, J., Urdaneta, E., Pascual, M., Nunez, M., Zudaire, E., & Santidrian, S. (1997). Impaired intestinal sugar transport in cirrhotic rats. *Gastroenterology*, *113*, 1180-1187.

[63] Ortega, M. A., Torres, M. I., Fernandez, M. I., Rios, A., Sanchez-Pozo, A., & Gil, A. (1997). Hepatotoxic agent thioacetamide induces biochemical and histological alterations in rat small intestine. *Dig. Dis. Sci.*, *42*, 1715-1723.

[64] Onori, P., Morini, S., Franchitto, A., Sferra, R., Alvaro, D., & Gaudio, E. (2000). Hepatic microvascular features in experimental cirrhosis. *J. Hepatol.*, *33*, 555-563.

[65] Ercan, N., Gannon, M. C., & Nuttall, F. Q. (1994). Liver glycogen synthase, phosphorylase, and the glycogen concentration in rats given a glucose load orally: a 24-hour study. *Arch. Biochem. Biophys.*, *315*, 35-40.

[66] Sasse, D. (1975). Dynamics of liver glycogen. The topochemistry of glycogen synthesis, glycogen content and glycogenolysis under the

experimental conditions of glycogen accumulation and deplection. *Histochemistry*, *45*, 237-254.

[67] Kuwajima, M., Golden, M. S., Katz, J., Unger, R. H., Foster, D. W., & McGarry, J. D. (1986). Active hepatic glycogen synthesis from gluconeogenic precursors despite high tissue levels of fructose 2,6-biphosphate. *J. Biol. Chem.*, *261*, 2632-2637.

[68] Gitzelmann, R., Steinmann, B., & Van den Berghe, G. (1989). Disorders of fructose metabolism. In C. R. Scriver, A. I. Beaudet, W. S. Sly, & D. Valle (Eds.), *Metabolic basis of inherited disease* (pp. 399-424). N.Y.: McGraw Hill.

[69] De Kalbermatten, N., Ravussin, E., Maeder, E., Geser, C., Jequier, E., & Felber, J. P. (1980). Comparison of glucose, fructose, sorbitol, and xylitol utilization in humans during insulin suppression. *Metabolism*, *29*, 62-67.

[70] Beringer, A., & Thaler, H. (1964). Uber quantitative untersuchungen des glykogengehaltes der leber bei gesunden und kranken menschen. *Wien. Klin. Wschr.*, *76*, 627-631.

[71] Nilsson, L. H., & Hultman, E. (1974). Liver and muscle glycogen in man after glucose and fructose infusion. *Scand. J. Clin. Lab. Invest.*, *33*, 5-10.

[72] Kruszynska, Y. T., Meyer-Alber, A., Darakshan, F., Home, P. D., & McIntyre, N. (1993). Metabolic handling of orally administrated glucose in cirrhosis. *J. Clin. Invest.*, *91*, 1057-1066.

[73] Bollen, M., Keppens, S., & Stalmans, W. (1998). Specific features of glycogen metabolism in liver. *Biochem. J.*, *336*, 19 –31.

[74] Ciudad, C. J., Massague, J., & Salavert, A. (1980). Synthesis of glycogen from fructose in the presence of elevated levels of glycogen phosphorylase in rat hepatocytes. *Mol. Cell. Biochem.*, *30*, 33-38.

[75] Kudryavtseva, M.V., Bezborodkina, N. N., & Kudryavtsev B. N. (1999). Restoration of the glycogen-forming function of hepatocytes in rats with liver cirrhosis is facilitated by a high-carbohydrate diet. *Br. J. Nutr.*, *81*, 473-480.

[76] Kudryavtseva, M.V., Bezborodkina, N. N., Okovity, S. V., & Kudryavtsev B. N. (2003). Effects of the 2-ethylthiobenzimidazole hydrobromide (bemithyl) on carbohydrate metabolism in cirrhotic rat liver. *Exp. Toxic. Pathol.*, *54*, 339-347.

[77] Hers, H.-G. (1990). Mechanisms of blood glucose homeostasis. *J. Inherit. Metab. Dis.*, *13*, 395-410.

[78] Hers, H. G. (1976). The control of glycogen metabolism in the liver. *Annu. Rev. Biochem.*, *45*, 167-189.

[79] Aiston, S., Andersen, B., & Agius, L. (2003). Glucose 6-Phosphate regulates hepatic glycogenolysis through inactivation of phosphorylase. *Diabetes*, *52*, 1333-1339.

[80] Magnusson, I., Rothman, D. L., Jucker, B., Cline, G. W., Shulman, R. G., & Shulman, G. I. (1994). Liver glycogen turnover in fed and fasted humans. *Am. J. Physiol.*, *266*, E796-E803.

[81] Ciudad, C. J., Carabaza, A., & Guinovart, J. J. (1988). Glycogen synthesis from glucose and fructose in hepatocytes from diabetic rats. *Arch. Biochem. Biophys.*, *267*, 437-447.

[82] Barrett, E. J., Bevilacqua, S., DeFronzo, R. A., & Ferrannini, E. (1994). Glycogen turnover during refeeding in the postabsorptive dog: implications for the estimation of glycogen formation using tracer methods. *Metabolism*, *43*, 285-292.

[83] Cadefau, J., Bollen, M., & Stalmans, W. (1997).Glucose-induced glycogenesis in the liver involves the glucose-6-phosphate-dependent dephosphorylation of glycogen synthase. *Biochem. J.*, *322*, 745-750.

[84] Agius, L., Peak, M., Newgard, C. B., Gomez-Foix, A. M., & Guinovart, J. J. (1996). Evidence for a role of glucose-induced translocation of glucokinase in the control of hepatic glycogen synthesis. *J. Biol. Chem.*, *271*, 30479 –30486.

[85] Niewoehner, C. B., Gilboe. D. P., & Nuttall. F. Q. (1984). Metabolic effects of oral glucose in the liver of fasted rats. *Am. J. Physiol.*, *246*, E89-E94.

[86] Niewoehner, C. B., & Nuttall, F. Q. (1988). Relationship of hepatic glucose uptake to intrahepatic glucose concentration in fasted rats after glucose load. *Diabetes*, *37*, 1559-1566.

[87] Newgard, C. B., Foster, D. W., & McGarry, J. D. (1984). Evidence for suppression of hepatic glucose-6-phosphatase with carbohydrate feeding. *Diabetes*, *33*, 192-195.

[88] Nordlie, R. C., Johnson, T. W., Cornatzer, W. E., & Twedell, G. W. (1979). Stimulation by polyamines of carbamylphosphate: glucose phosphohydrolase activities of multifunctional glucose-6-phosphatase. *Biochem. Biophys. Acta*, *585*, 2-23.

[89] Belfiore, F., Romeo, F., Iannello, S., & Salamone, C. (1989). The glucose-6-phosphatase/glucokinase ratio in the liver of obese-diabetic subjects. *Biochem. Med. Metab. Biol.*, *41*, 77-80.

[90] Jungermann, K., & Katz, N. (1986). Metabolism of carbohydrates. *In Regulation of hepatic metabolism* (pp. 211-232). New York: Plenum Press.

[91] Huang, Y. T., Hsu, I. S., Chen, C. J., Liu, C. T., & Wie, Y. H. (2003). Oxidative-stress-related changes in the livers of bile-duct-ligated rats. *J. Biomed. Sci.*, *10*, 170-178.

[92] Proctor, E., & Chatamra, K. (1982). High yield micronodular cirrhosis in the rat. *Gastroenterology*, *83*, 1183-1190.

[93] Fischer-Nielsen, A., Poulsen, H. E., Hansen, B. A., Hage, E., & Keiding, S. (1991). CCl4 cirrhosis in rats: irreversible histological changes and differentiated functional impairment. *J. Hepatol.*, *12*, 110-117.

[94] Planaguma, A., Claria, J., Miquel, R., Lopez-Parra, M., Titos, E., Masferrer, J. L., & Rodes, J. (2005). The selective cyclooxygenase-2 inhibitor SC-236 reduces liver fibrosis by mechanisms involving non-parenchymal cell apoptosis and PPARgamma activation. *FASEB J.*, *19*, 1120-1122.

[95] Kudryavtseva, M. V., Bezborodkin, N. N., Okovity, S. V., & Kudryavtsev, B. N. (2001). Metabolic heterogeneity of glycogen in hepatocytes of patients with liver cirrhosis: the glycogen of the liver lobule zones in cirrhosis. *Eur. J. Gastroenterol. Hepatol.*, *13*, 693-697.

[96] Pieniazek, D., Pronicka, E., Cabalska, B., Borowska, B., Miloszewska, E., Maciejko, D., Kulczycka, H., & Prokopowicz, K. (1985). Glycogenosis type VI and VIa. *Pediat. Pol.*, *60*, 489-496.

[97] Kudryavtseva, M. V., Emelyanov, A. V., Sakuta, G. A., Bezborodkina, N. N., & Kudryavtsev, B. N. (1998). Glycogen-forming function of hepatocytes in rat regenerating cirrhotic liver after a partial hepatectomy. *Tissue Cell.*, *30*, 200-206.

[98] Kudryavtseva, M. V., Emelyanov, A. V., Sakuta, G. A., Scorina, A. D., Sleptsova, L. A., & Kudryavtsev, B. N. (1992). Cytofluorimetric study of the content of glycogen and its fraction in hepatocytes of patients with liver cirrhosis of different etiologies. *Tsitologiya*, *34*, 100-107.

[99] Francavilla, A., Jones, A. F., & Starzl, T. E. (1978). Cyclic AMP metabolism and adenylate cyclase concentration in patients with advanced hepatic cirrhosis. *Gastroenterology*, *75*, 1026-1032.

[100] Rosenfeld, E. A., & Popova, I. A. (1989). *Inborn errors of glycogen metabolism*. Moscow, Russia: Meditsina.

[101] Henderson, J. M., Scott, S. S., Merrill, A. H., Hollins, B., & Kutner, M. H. (1989). Vitamin B6 repletion in cirrhosis with oral pyridoxine. *Hepatology*, *9*, 582-588.

[102] van Dijk, T. H., van der Sluijs, F. H., Wiegman, C. H., Baller, J. F., Gustafson, L. A., Burger, H. J., Herling, A. W., Kuipers, F., Meijer, A. J. & Reijngoud, D. J. (2001). Acute inhibition of hepatic glucose-6-phosphatase does not affect gluconeogenesis but directs gluconeogenic flux toward glycogen in fasted rats. A pharmacological study with the chlorogenic acid derivative S4048. *J. Biol. Chem*, *276*, 25727-25735.

[103] Shimamura, J. (1987). Hepatic key carbohydrate metabolizing enzymes in bening liver disease and hepatocellular carcinoma. *J. Med. Assos. Okayama*, *99*, 261-272.

[104] Benedetty, A., Casini, A., Ferrali, M., & Comporti, M. (1977). Studies on the relationships between carbon tetrachloride-induced alterations of liver microsomal lipids and impairment of glucose-6-phosphatase activity. *Exp. Mol. Pathol.*, *27*, 309-323.

[105] De Groot, H., Noll, T., & Toll, T. (1985). Loss of latent activity of liver microsomal membrane enzymes evoked by lipide peroxidation. Studies of nucleoside diphosphatase, glucose-6-phosphatase, and UDP glucuronyltransferase. *Biochem. Biophys. Acta*, *815*, 91-96.

[106] Ohyashiki, T., Kamata, K., Takeuchi, M., & Matsui, K. (1995). Contribution of peroxidation products to oxidative inactivation of rat liver microsomal glucose-6-phosphatase. *J. Biochem.*, *118*, 508-514.

In: Glycogen ISBN: 978-1-62948-395-5
Editors: P.L. Weiss and B.D. Faulkner

Chapter 3

Muscle and Liver Glycogen in Helminth-Infected Fish

Galina I. Izvekova** *and Evgeny I. Izvekov

Papanin Institute for Biology of Inland Waters, Russian Academy of Sciences, Borok, Nekouzskii Raion, Yaroslavskaya Oblast, Russia

Abstract

The influence of trematode and cestode infections on the glycogen content in the muscles and liver of various fish species was studied. The glycogen content was found to increase in the muscles of young cyprinid fishes (ide *Leuciscus idus*, dace *Leuciscus leuciscus*, goldfish *Carassius auratus*, and roach *Rutilus rutilus*) following the experimental infection with the metacercariae of the trematode *Apophallus muehlingi*. The same was true for the roach juveniles naturally infected with another trematode *Posthodiplostomum cuticola*. These changes could be explained by a decrease in swimming activity of fish due to parasite-induced muscle injuries. As an example, a significant drop in spontaneous motor activity was shown for the crucian carp *Carassius carassius* after 30-day maintenance with the trematode-shedding mollusks, which may be a possible reason for glycogen accumulation in the muscles of infected fish.

At the same time, glycogen concentrations in the muscles and liver declined when the fish (perch *Perca fluviatilis*, bream *Abramis brama*, and sabrefish *Pelecus cultratus*) were infected with the plerocercoids of

* Corresponding author: izvekov@ibiw.yaroslavl.ru.

some cestodes (*Triaenophorus nodulosus* and *Ligula intestinalis*). Reduced glycogen content in the liver testifies to a dysfunction of this organ in the infected fish and, as a consequence, leads to a decrease in the activity of major digestive enzymes (proteinases and glycosidases) in the intestine. Thus, the infection with trematodes and cestodes differently affects the glycogen content in the fish tissues and, respectively, might cause different consequences for animal health.

INTRODUCTION

Glycogen is known to be the main reserve polysaccharide and a key energy source for fishes and other animals (Hochachka, 1985; Weber, 1992; Weber et al., 1996). In an organism, the glycogen content serves as an important indicator of vital activity. The glycogen level in organs is changeable: it is spent as required and stored up again at extra glucose intake with food (Scarabello et al., 1991). The studies on glycogen content in various organs of fish are carried out for a long time. There exist comprehensive reviews synthesizing the available data (Plisetskaya & Kuzmina, 1972; Plisetskaya, 1975; Thillart & Raaij, 1995), as well as articles regarding the glycogen content as an additional indicator of fish physiological state (Ottolenghi et al., 1981; Thillart & Smit, 1984; Scarabello et al., 1991). In fishes, the greatest amounts of glycogen are stored in liver, brain, and red muscles (Plisetskaya, 1975). It should be noted that in most fishes, due to deep interpenetration of pancreas and liver, these organs are collectively considered as hepatopancreas, though their cells are microscopically differentiated and independent of each other (Sorvachev, 1982).

A great body of literature is devoted to the glycogen content in different organs as a function of various factors, such as diet, swimming activity, season, etc. (Plisetskaya & Kuzmina, 1972; Johnston & Goldspink, 1973 a, b; Johnston et al., 1977; Ottolenghi et al., 1981; Thillart & Smit, 1984; Scarabello et al., 1991; Coban & Sen, 2011; Liew et al., 2012, etc.). For instance, the liver glycogen in some fish decreases at high stocking density in ponds (Trenzado et al., 2008). The muscle glycogen in fish depends on their age, and, especially, on the muscular functional load (Thillart & Smit, 1984; Parkhouse et al., 1988 a, b; Weber & Haman, 1996). It is known that glycogen reserves in fish muscles are actively spent during the intense exercise stress related to migrations (Plisetskaya, 1975).

Studying the host–parasite interactions is one of the basic trends in parasitology. Within this framework, the focus is set on the influence of different parasitic infections on the biochemical parameters of a host organism, including the glycogen levels (Thompson, 1983, 1997). Quite often the data obtained are inconsistent. For example, in various animals parasitized by certain species of helminths, glycogen depletion was observed. Thus, in rabbit, guinea pig and some other hosts infected with *Fasciola hepatica*, the liver glycogen decreased (Kublickiene & Goldbergiere, 1966), as well as in golden hamsters experimentally infected with plerocercoids of *Spirometra erinacei* (Tsuboi et al., 1991). At the same time, in *Trichinella*-infected mice, the blood sugar was reduced, while the muscle glycogen increased markedly, with insignificant changes in liver glycogen. However, in the mice infected with another nematode *Nematospiroides dubius*, a heavy decrease in liver glycogen was found, attributable to a reduced food intake by parasitized animals (Symons & Jones, 1974).

The information on the influence of helminthic infection on the glycogen content in their host fishes is scarce and fragmentary. Therefore we have made an attempt here to synthesize our own long-term experimental data (Izvekova & Kuz'mina, 1996; Kuperman et al., 1997; Izvekova 2001 b; Izvekova & Tyutin, 2011 a, b; Izvekov et al., 2012; Tyutin & Izvekova, 2013). These data were obtained during the laboratory and field studies with several fish species inhabiting the Rybinsk reservoir (Russian Federation, 58°22´30´N 38°28´040´E). We concentrate on the changes in muscle and liver glycogen of fishes parasitized by some trematodes and cestodes, as well on the consequences of these changes for the host vital functions.

1. Influence of Trematode Infection on Muscle Glycogen and Motor Activity in Young Cyprinids

1.1. Glycogen Content

Trematoda is an extensive class of flat worms having a complex life cycle involving more than one host. Representatives of this class are causative agents of many severe diseases of humans and animals. A bulk of publications is devoted to the effects of infection with trematode parthenitae on biochemical composition of their first intermediate molluskan hosts. Thus, in the tissues of parasitized individuals of some freshwater gastropods, a decrease in glycogen and amino acid levels was noted, along with changes in fatty-acid composition (see for reviews Thompson, 1983, 1997).

Glycogen depletion in the digestive gland of infected *Littorina saxatilis* revealed by means of thin-layer chromatography testifies to functional abnormalities affecting its synthesis in the presence of trematode parthenitae (Arakelova et al., 2003, 2004). In trematode-infected snails, alterations of carbohydrate and lipid metabolism were reported, as well as changes in protein spectrum and content of protein sulfhydryl groups in the hemolymph (Stadnichenko, 1978; Pinheiro et al., 2009). The degree of manifestation of these changes depends on the localization of parasites and invasion intensity.

The least understood issue is still the influence of trematode metacercariae on their second intermediate hosts, particularly fishes. Solitary studies are devoted to this problem. For example, in the muscles and liver of the round goby *Neogobius melanostomus* heavily infected with *Cryptocotyle concavum* metacercariae, lowered levels of total lipids and triglycerides were noted, as well as a decrease in liver cholesterol (Shchepkina, 1981). In juvenile bluegill *Lepomis macrochirus* parasitized by the trematodes *Uvulifer ambloplitis*, the total lipid content was reduced both in the laboratory and field environment (Lemly & Esch, 1984).

As fishes, particularly juveniles, often serve as second intermediate hosts for the trematodes, the data relevant to the fluke infections on the fish organism are of principal importance. Thus, the metacercariae of trematodes *Apophallus muehlingi* and *Posthodiplostomum cuticola* cause the "black spot disease" resulting in elevated mortality of young cyprinids at high invasion intensities. Most commonly the metacercariae are localized on the fin rays, in the intercostal and tail muscles, and subcutaneously. Piscivorous birds are known to be the definitive hosts of these flukes (Odening, 1970; Biserova, 1990; Izvekova & Tyutin, 2011 b).

In our studies with experimental infection (Izvekova & Tyutin, 2011 b; Tyutin & Izvekova, 2013), young-of-the-year specimens of the following fishes were used: goldfish *Carassius auratus gibelio*, roach *Rutilus rutilus,* ide *Leuciscus idus*, and dace *Leuciscus leuciscus*. In the laboratory, the effect of parasitism by *Apophallus muehlingi* metacercariae on the muscle glycogen of the young fish was examined. Two samples of each species (5 to 10 individuals each) were placed in two identical aquariums. Each aquarium contained either ten mollusks *Lithoglyphus naticoides* infected with *A. muehlingi* parthenitae (experiment) or ten unparasitized mollusks (control). Additionally, the effect of natural infection with *Posthodiplostomum cuticola* on the muscle glycogen was studied in the young roach from the Rybinsk reservoir. The glycogen concentrations were determined by the Montgomery method (Montgomery, 1957) and expressed as mg per g of wet tissue.

Absorbance was measured using the spectrophotometer Lambda 25 at the wave length of 490 nm.

Our laboratory tests showed that the young cyprinids infected with *Apophallus muehlingi* metacercariae had significantly higher ($p < 0.05$) muscle glycogen levels than their uninfected counterparts (Table 1). The studies with juvenile roach made it possible to compare the impact of natural infection with *Posthodiplostomum cuticola* metacercariae and three-day experimental infection with *Posthodiplostomum cuticola* metacercariae. It was shown that muscle glycogen tends to increase both in the naturally and experimentally infected roach (Table 1).

Table 1. Glycogen content (mg/g) in the muscles of uninfected fishes (control) and those infected with trematode metacercariae (test)

Fish species	Trematode species	Test duration, days	Control	Test
Ide	*A. muehlingi*	30	0.48 ± 0.04* 269 ± 27*	0.79 ± 0.06* 497 ± 18*
Dace	*A. muehlingi*	30	0.76 ± 0.05* 648 ± 35	1.09 ± 0.05* 574 ± 23
Goldfish	*A. muehlingi*	3	2.36 ± 0.13 243 ± 6.27	2.86 ± 0.34 245 ± 15
Goldfish	*A. muehlingi*	7	0.94 ± 0.08* 916 ± 79	1.26 ± 0.08* 856 ± 73
Roach	*A. muehlingi*	3	3.14 ± 0.25* 728 ± 60*	5.3 8± 0.68* 946 ± 67*
Roach	*P. cuticola*	natural infection	3.04 ± 0.30 517 ± 26*	3.98 ± 0.42 709 ± 35*

Mean ± SE indicated. Asterisks show significant differences between the test and control samples ($p < 0.05$).

During the experiments, up to several hundred trematode larvae penetrated into the muscules of each fish, nonetheless no host mortality occurred. Thus, the relations within the studied "metacercariae-cyprinid fish" system revealed no signs of pronounced antagonism and narrow species specificity as well.

It was also found that in the muscles of the juvenile goldfish and dace infected with *Apophallus muehlingi* metacercaria, the protein content hardly changes, but significantly increases in the ide and roach (Izvekova & Tyutin, 2011 b; Tyutin & Izvekova, 2013) (Table 1). The glycogen and protein levels in the muscles of the young-of-the-year fish are species-dependent. These levels are extremely variable and account not only for trematode infection but also for environmental conditions (in our case, natural or experimental infection, different test duration etc.). It was shown earlier that in artificially raised juveniles of the carp *Cyprinus carpio*, mixed invasion with protists, monogeneans and cestodes causes an increase in protein level in blood serum, heart, kidneys, liver, spleen, intestine, and muscles (Kurovskaya, 2002). In this context, the lack of helminth influence on the protein content in the muscles of goldfish and dace in our experiments can be explained by their higher resistance to parasitic infection in comparison with ide and roach.

The increased muscle glycogen content in our infected fish may result from several factors. One of them could be the lower motor activity of parasitized individuals. The glycogen reserves are reported to be spent intensively during short-term bursts of activity in vertebrates, including fish (Moyes & West, 1995; Lowe & Wells, 1997). When parasitizing their intermediate or supplementary hosts, the helminth larvae are known to alter host behavior, which facilitates their reaching a definitive predatory host (see for a review Barber et al., 2000; Barber & Wright, 2006). Apparently, the penetration of trematode larvae into the fish causes muscle injuries affecting the host motor activity. The reduced activity of the infected fish can increase their susceptibility to predation, enhancing the parasite's chances to complete its life cycle. Therefore, in the natural environment, one could expect higher selective predation pressure and elevated elimination rates for the juveniles of those species that are most susceptible to metacercarial infection.

The changes in the endocrine status of the parasitized fish may represent another reason for the elevated muscle glycogen. The carbohydrates entering the fish organism with food are stored in the muscles as glycogen, with the storage hormone, insulin, playing a key role in this process. The high endocrine activity may result in glycogen accumulation in the muscles. The alterations in physiological status owing to living conditions affect the endocrine composition of the blood (i.e., the catecholamine and insulin levels) and the carbohydrate metabolism as well (Plisetskaya & Kuzmina, 1972). The muscle glycogen content in some teleosts was reported to increase following the insulin injections. In some cases, corticosteroids, i.e. hormones promoting the glycogen synthesis from non-carbohydrate sources, may favor glycogen accumulation in tissues (Plisetskaya, 1975).

Thus, all the cyprinids in our studies exhibited a considerable increase in the muscle glycogen level in response to the infestation with trematode metacercariae. It is noteworthy that an increase in glycogen content was also observed in the abdomens of crustaceans *Gammarus insensibilis* parasitisized by the salt marsh trematode *Microphallus papillorobustus* (Ponton et al., 2005). This similarity may illustrate a common tendency in metabolic changes caused by the avian trematodes in phylogenetically distant second intermediate hosts. On the other hand, elevated glycogen levels were also found in crustaceans infected by other parasites (Amat et al., 1991; Plaistow et al., 2001), suggesting that this pattern may not be restricted solely to trematode parasitism. Further comparative studies with representatives of different host taxa are needed to clarify this issue.

1.2. Swimming Activity

Glycogen represents one of the main energy sources available for motor activity, and its content in the muscles is indicative of the vital functions in parasitized fish. Currently, the impact of trematode infection on swimming behavior of fishes as second intermediate hosts is studied insufficiently. A few relevant publications refer either to non-cyprinid fish (Santos et al., 2011) or to a short-term motor response of the fish to cercareae penetration (Baldwin et al., 1967; Brassard et al., 1982).

Therefore we performed a case study of swimming behavior with the young-of-the-year crucian carp *Carassius carassius* parasitized by the metacercariae of *Apophallus muehlingi* (Izvekov et al., 2012). The studies were carried out with four samples of fish (15 specimens in each): two control groups, and two groups of experimentally infected fish differing in their parasitic load. For this purpose, each sample was preliminarily kept in a separate aquarium together with ten *Lithoglyphus naticoides* specimens. Experimental aquariums contained the cercariae-shedding snails, with uninfected snails placed in the control aquariums. The experimental and control fish were maintained with the snails for 7 days to gain the lighter parasitic load, and for 30 days to ensure the heavier load.

For activity observations, the fish were individually placed into a uniformly lit annular swimway (with external diameter of 24 cm and corridor width of 4.5 cm). The swimming behavior was recorded with a digital video camera for a period of 20 minutes. The video records were analyzed using "FishTracker" software developed in the Institute for Biology of Inland Waters RAS and Schmidt Institute of Physics of the Earth RAS. The program calculated the angular distance (degrees) travelled by the fish. On termination of the experiment, encysted metacercareae in the whole fish organism were counted. The mean ± SE number of cysts after the 30-day exposure (18.1 ± 1.8; 10 to 33 cysts) significantly exceeded that after the 7-day exposure (12.4 ± 1.4; 5 to 23 cysts) ($p < 0.05$).

Following the 7-day stay with infected mollusks, the crucian carps exhibited no evident changes in their activity level (Figure). The mean distance travelled by the infected fish for the 20-minute period was slightly higher compared to controls, but this 5% difference was insignificant ($p > 0.05$). However, the temporal patterns of activity were different in parasitized and parasite-free fish. Thus, the control fish quickly became familiar with their new environment and then supported a relatively constant activity level through the test. On the contrary, in the weakly infected group, the activity level tended to increase over the test (linear regression, $R^2 = 0.325$; $p = 0.0087$). A local maximum of activity was observed in 10 to 13 minutes after the start of the experiment (23–30% higher as compared to control, $p >$ 0.05).

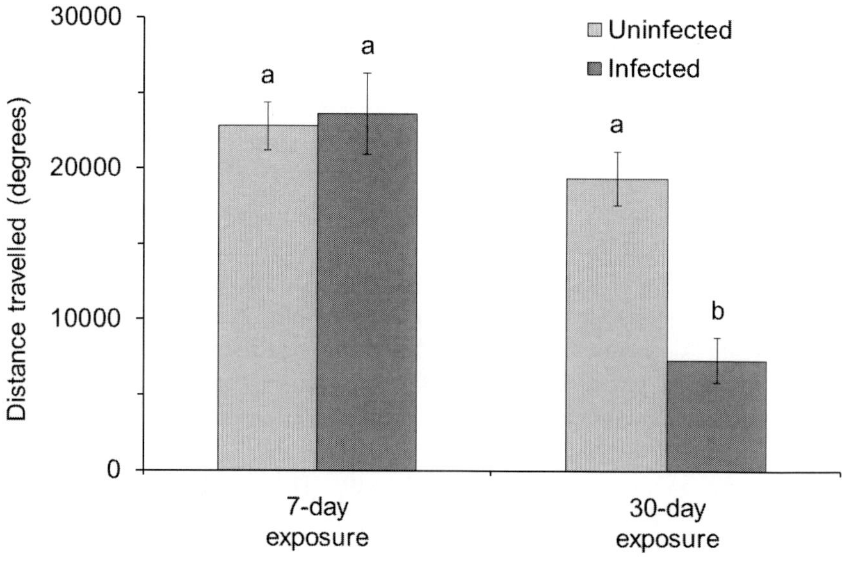

Figure 1. Motor activity of juvenile crucian carps *Carassius auratus* in the annular swimway after their joint keeping with the snails *Lithoglyphus naticoides* parasitized (or unparasitized) by *Apophallus muehlingi*. Means ± SE for the 20-minute observation period are given. Means with different letters are significantly different ($p < 0.0001$).

In contrast, keeping the crucian carps with the infected mollusks for a month caused a drastic change in the fish behavior (Figure 1). The motor activity of the heavily infected fish significantly decreased compared to controls (by 164% for the 20-minute period; $p = 0.00002$). Similarly to the short-term exposure scenario, the activity of heavily infected crucian carps gradually increased towards the end of the observation period (linear regression, $R^2 = 0.671$; $p = 0.00001$). Most individuals exhibited intermittent swimming, frequently stopping and rising to the surface, which would potentially make them more visible to definitive hosts in the wild.

Currently, numerous examples of altered swimming behavior in trematode-infected fish are known (see for reviews Barber et al., 2000; Barber & Wright, 2006). However, most reports concern fish stamina (Klein et al., 1969; Butler & Millemann, 1971; Sogandares-Bernal et al., 1979; Lutterschmidt et al., 2007) or avoidance response (Blake et al., 2006). Meanwhile, the information on spontaneous motor activity seems most relevant to the above-mentioned effect of muscle glycogen accumulation. In this context, our behavioral data are in agreement with those for some other fishes, particularly for *Poecilia vivipara* parasitized by the trematode *Acanthocollaritrema umbilicatum* (Santos et al., 2011). Metacercarial cysts of this fluke occur among the scales, fins, and musculature and in the buccal cavity of naturally infected *P. vivipara* (Simões et al., 2008). When tested in 17–27 days after experimental infection, the specimens of *P. vivipara* harboring moderate numbers of metacercareae displayed enhanced motor activity while the heavily infected fish reduced their activity level (Santos et al., 2011). These data suggest the ability of the parasites to modify the swimming behavior of the host fish.

On the other hand, in some studies, a decrease in swimming activity of the trematode-infected fish was observed shortly after the experimental infection. Thus, guppies *Poecilia reticulata* exposed to 25–100 cercareae of the trematode *Diplostomum spathaceum* showed a reduced activity within a 70-min post-exposure period, which made them more susceptible to predation by brook trout *Salvelinus fontinalis* (Brassard et al., 1982). Similarly, the juvenile salmonids infected with the trematode *Nanophyetus salmonicola* decreased their motility 1–2 hr after exposure to infected snails *Oxytrema silicula* and displayed no characteristic darting movements (Baldwin et al., 1967). These behavioral changes are considered a result of injuries caused by the cercariae migrating through the host tissues. The list of injuries includes destruction of muscle tissue, extensive subcutaneous hemorrhages at the fin bases, orbital hemorrhage, as well as partial renal tubule and blood vessel occlusion (Baldwin et al., 1967). As in our experiments crucian carps might be attacked by cercareae round the whole exposure period, the continuing tissue damage might contribute to the decrease in the motor activity of the fish.

Thus, through the example of crucian carp, a significant decrease in spontaneous motor activity was shown for cyprinid fish after their long contact with trematode-infected mollusks, which may be a possible reason for the glycogen accumulation in the muscles of parasitized cyprinids.

2. Influence of Cestodes Infection on the Muscle and Liver Glycogen in Fishes

The Cestoda class includes about 3500 species of parasitic tapeworms living mainly in the organism of vertebrates. Many of these flatworm species parasitize fish, using them as intermediate or definitive hosts.

2.1. *Ligula intestinalis*

Plerocercoid larvae of *Ligula intestinalis* (L.) living in the body cavity of fish are widespread parasites of many cyprinids, causing ligulosis epizooties in some species (Baruš et al., 1997; Shields et al., 2002). As the tapeworms grow and develop, they exert both mechanical and toxic effects on the fish organism (Dubinina, 1980). *Ligula* plerocercoids negatively influence the blood picture, growth rate and condition, fat content and reproductive system of cyprinids (Arme & Owen, 1968; Markov et al., 1978; Soutter et al., 1980; Taylor & Hoole, 1989). Currently, a lot of attention is paid to the parasite's distribution and systematics (Matskási & Juhász, 1977; Luo et al., 2003), its molecular characterization (Li et al., 2000), activity of the enzymes involved in the intermediary carbohydrate metabolism (McManus, 1975; McManus & Sterry, 1982; Smyth & McManus, 1989), including its final stages (Barrett, 1978; Izvekova, 2001 a). A lot of studies are devoted to protease inhibitor activities (Matskási & Németh, 1979), relationships of the parasite with intermediate hosts and its influence on the host vital functions (Arme & Owen, 1970; Taylor & Hoole, 1989; Izvekova & Kuz'mina, 1996; Vanacker et al., 2012).

To a large extent, the interest in studying plerocercoids is explained by the fact that the environmental specifics of a particular water body influence the host fishes, which has an effect on the interactions between the plerocercoids and their intermediate hosts, and finally, on the outcome of a parasitological research. Therefore many publications focus on those details of such interactions that are characteristic of the given water bodies and particular host species. Piscivirous birds serve as definitive hosts for *L. intestinalis* (Dubinina, 1980).

In the Rybinsk reservoir, the plerocercoids of *L. intestinalis* occur in various cyprinids (Kuperman et al., 1997). The most common second intermediate host for plerocercoids is the bream (Kuperman et al., 1997), while the sabrefish is the most rarely infected species (Izyumova, 1987). Therefore these two fish species were selected to study the influence of infection with *L. intestinalis* plerocercoids on the muscle and liver glycogen.

The glycogen content was determined by the Montgomery method (Montgomery, 1957) and represented as mg per g of wet tissue. Additionally, we analyzed the activity of glycosidases (total activity of α-amylase EC 3.2.1.1, glucoamylase EC 3.2.1.3, and maltase EC 3.2.1.20) by Nelson's method (Ugolev et al., 1969) from the increase in hexose level. Besides, the total proteolytic activity (activities of trypsin EC 3.4.21.4, chemotrypsin EC 3.4.21.1, and dipeptidases EC 3.4.13.18) was determined by Anson's method (Anson, 1938) based on the liberation of tyrosine. The activity of glycosidases was determined using 1.8% (w/v) soluble starch as the substrate (pH 7.4). The total activity of proteinases was measured using 1% (w/v) casein solution as the substrate (pH 8.0). Enzyme activities were expressed in μmol/g/min. The optical densities were measured with the use of the spectrophotometer Lambda 25 at the wave length of 490 nm (glycogen), 560 nm (glycosidases), and 760 nm (proteinases).

2.1.1. Ligula intestinalis – Bream

The glycogen content was determined in the muscles and hepatopancreas of unparasitized and ligulosed breams (4 5 years old and 16–18 cm long) (Table 2). *Ligula* infection was found to reduce the glycogen content both in the muscles and liver ($p < 0.05$). Parasitism by *L. intestinalis* had a greater effect on the glycogen reserves in muscles of fish than in their liver. For example, in ligulosed specimens, a 57% decrease in muscle glycogen was recorded versus a 44% decrease in the liver. Also, the L/R ratio, i.e., liver glycogen content (L) divided by the muscle glycogen (M) was altered following the infection (Table 2).

Table 2. Muscle and liver glycogen content (mg/g), and the L/R ratio in the uninfected fish (U) and infected (I) with pleroceroid *L. intestinalis*

Fish species	Muscles		Liver		L/R ratio	
	U	I	U	I	U	I
Bream	1.5 ± 0.2	0.70 ± 0.30*	46 ± 2	26 ± 7*	30.7	37.1
Sabrefish	0.41 ± 0.04	0.33 ± 0.09	0.45 ± 0.06	0.56 ± 0.06	1.1	1.7

Mean ± SE indicated. Asterisks show significant differences between the test and control samples ($p < 0.05$).

Besides, the analysis of biochemical composition of the parasite-free and ligulosed breams showed that the infected fish had greater water content and lower protein level, while the total lipids were practically unaffected (Izvekova & Kuz'mina, 1996; Kuperman et al., 1997).

Our studies also demonstrated a considerable decrease in the performance of the enzymes providing the initial stages of assimilation of protein and carbohydrate food components in the *Ligula*-infected bream (Table 3). Due to the infection, the activities of proteases and glycosidases in bream dropped by 33% and 24%, respectively. Based on the data obtained, the G/P ratio (i.e. glycosidase activity divided by the protease activity) was calculated. In parasitized breams, this ratio (G/P = 1.90) appears to be higher compared to controls (G/P = 1.68). This implies that *Ligula* infection causes changes in relative activity of these two enzyme groups, and thereby, feeding efficiency of fish may decrease due to the reduced enzyme activity. Besides, in case of invertase, considerable variations in kinetic properties of the enzyme accomplishing the membrane digestion were shown for the ligulosed breams (Izvekova & Kuz'mina, 1996).

Table 3. The activity (μmol/g/min) of proteinases (PA) and glycosidases (GA) in the intestinal mucosa of the fish uninfected (U) and infected (I) with pleroceroid *L. intestinalis*

Fish species	PA		GA	
	U	I	U	I
Bream	16.4 ± 2.2	11.0 ± 1.3*	27.5 ± 2.5	20.9 ± 2.9*
Sabrefish	2.2 ± 0.3	1.3 ± 0.2	1.2 ± 0.2	0.6 ± 0.1*

Mean ± SE indicated. Asterisks show significant differences between the test and control samples ($p < 0.05$).

2.1.2. Ligula intestinalis – Sabrefish

The sabrefish *Pelecus cultratus* has been reported as another cyprinid host of *L. intestinalis* (Dubinina, 1980), but the information on this host–parasite pair is almost absent (Izvekova & Tyutin, 2011 a).

To meet the lack of information, we determined the glycogen content in the liver and muscles of infected and uninfected sabrefish yearlings (12–14 cm long) (Table 2). It was shown that the glycogen levels in both tested tissues of this intermediate host insignificantly changed in the ligulosed specimens ($p > 0.05$). At the same time, there were certain changes in the L/R ratio describing the relative concentration of glycogen in two tissues studied (Table 2). As well as in the bream, the L/R ratio in the sabrefish increased following infection. This ratio in the sabrefish was much lower compared to the bream, but the trend of these changes was common in both species, which may evidence the similarity of the processes caused by the plerocercoid parasitism.

The data on the glycogen level in the sabrefish tissues can be interpreted in different ways. On the one hand, the difference in tissue glycogen between the parasitized and parasite-free sabrefish is insignificant; on the other hand, the effect of infection on the carbohydrate metabolism manifests itself as changes in the L/R ratio. The L/R ratio increases following the infection, testifying to a glycogen imbalance in the tested tissues and to a decrease in relative glycogen content in the fish muscles.

As in the case of bream, the sabrefish harboring the plerocercoids displayed lower activity of the enzymes accomplishing the initial stages of protein and carbohydrate hydrolysis (Table 3). In the infected sabrefish, the enzyme activity decreased by 41% for proteinases and by 50% for glycosidases. The relative activity of glycosidases and proteinases expressed as the G/P ratio differed marginally in the infected and uninfected fish (0.49 and 0.54 respectively).

The muscle glycogen in the parasite-free bream was 3.7 times greater compared to the parasite-free sabrefish, while in the case of infected fish the interspecies differences appeared to be less pronounced (2.1 times). Even more profound contrast was found in the liver glycogen: when comparing the uninfected specimens, the bream had 102 times higher glycogen content than the sabrefish, whereas in the parasitized individuals the difference did not exceed 46 times. The considerable difference between these species in the glycogen content, particularly in the liver, can be explained by different levels of their energy metabolism related to diverse habit of life. The bream inhabits near-bottom areas, while the sabrefish is an active pelagic swimmer

(Reshetnikov, 2010). It is also noteworthy that the plerocercoid infection had a lesser effect on the sabrefish glycogen reserves as compared to the bream. This fact can be attributed to a higher resistance of the sabrefish to parasitic infection, which is corroborated by rare occurrence of ligulosed specimens in this species (Izyumova, 1987). Thus, in the Rybinsk reservoir, the relatively high prevalence of plerocercoid infection in the sabrefish was first reported in 2005 (Izvekova & Tyutin, 2011 a). The decrease in the muscle glycogen can be explained by the alterations in behavior and motor activity of the infected individuals. It is known that parasitized fish often display increased energy requirements. As a consequence, parasite-induced changes in host appetite, habitat selection, and time budget might greatly affect their motor activity. These behavioral changes are believed to facilitate the plerocercoid penetration into the definitive host organism (piscivorous bird) (Barber et al., 2000; Loot et al., 2001; Barber, 2003).

In this regard, it is worth mentioning that the activity of digestive hydrolases also differed in the bream and sabrefish. Thus, the proteolytic activity in bream was 7.5 times higher than that in the sabrefish when comparing the uninfected fish, and 8.5 higher in case of the infected specimens. The difference in the glycosidase activity was even more pronounced (22.9 and 34.8 times for unparasitized and parasitized fish, correspondingly). The differences between the two tested fishes in the activity of their digestive enzymes and glycogen levels probably account for the species specifity and feeding habits (the bream is a benthivore, while the sabrefish is a piscivore and facultative planktivore), as well as for the factor of age (four to five-year-old bream versus yearling sabrefish).

Interspecies differences in the impact of parasitic infections on various fish species were also noted by other authors. For instance, in the gudgeon (*Gobio gobio*), the infestation with plerocercoid *L. intestinalis* caused much less pronounced gonadal retardation and tissue proliferation in the body cavity compared to several other cyprinids (Arme & Owen, 1968).

A 2.8–3.7-fold decrease in the liver glycogen in ligulosed bream and roach as compared to uninfected controls was also reported by Kosareva (1961). In addition, it was shown that harboring *L. intestinalis* tended to depress both size and weight of the fish liver, which seems to be a result of reduced metabolism of fats and proteins. Concurrently, marked changes were detected in the blood parameters and serum protein levels, obviously representing part of the immune response of the fish (Arme, Owen, 1968; Sweeting, 1977).

Ligula infection leads to a decrease in the growth rate, condition and fecundity of fish (Kurovskaya, 1993), affecting their amino-acid and lipid status (Sidorov et al., 1989). According to some reports (Soutter et al., 1980), parasitization by *L. intestinalis* influences the fish in the way similar to starvation: for example, both tend to reduce the tissue glycogen content. At the same time, despite profound changes in vital functions, harboring *L. intestinalis* is not always fatal to its host fish where plerocercoids can live for 3 years and more (Dubinina, 1980). Furthermore, in some cases, plerocercoids had no effect on host fish health, with insignificant difference in condition values between infected and uninfected specimens (Vanacker et al., 2012).

Our data show a considerable decrease in activity of the enzymes responsible for the initial stages of hydrolyzing the major nutrients (proteins and carbohydrates), which entails lower efficiency of food assimilation. Most likely, the mechanical irritation caused by the parasite, along with its metabolites indirectly influences the activity of host digestive enzymes.

It should also be noted that the rates of synthesis and degradation of glycogen in the animal tissues are regulated hormonally, and both can be affected during various endocrine disorders. A typical response of the vertebrate organism to a stressor is associated with the release of adrenaline and noradrenaline (Rosen, 1984).

Infection with *Ligula* plerocercoids can be regarded as a kind of stressor agent (Richards & Arme, 1981). In this respect, altered behavior of the infected fish involves changes in their food spectra and, finally, a decrease in fish condition. The latter may be explained by the loss of nutrients that are required not only for parasite growth, but also for keeping the host homeostasis under the stress influence due to infestation (Richards & Arme, 1981). A simultaneous decrease in muscle and liver glycogen, along with lowered activity of digestive hydrolases in the infected fish suggests a contribution of stress hormones, including adrenaline, to these processes. It is known that adrenaline provokes the muscle and liver glycogen depletion and reduces the liver α-amylase activity (Plisetskaya, 1975). In addition, pancreatic cells of the hepatopancreas release inactive proteolytic enzymes (trypsinogen, chymotrypsinogen, and procarboxypolypeptidases A and B) into the intestine, as well as amylase, lipase and nuclease in their active forms (Sorvachev, 1982). Dysfunction of the hepatopancreas in the ligulosed fish can suppress synthesis of proteolityc zymogens and amylase, or their intake into the intestines.

Thus, lowered glycogen levels were observed in the muscles and liver of fish harboring *Ligula* plerocercoids. Moreover, the activity of digestive enzymes providing the initial stages of protein and carbohydrate hydrolysis decreased. The intensities of these alterations differed in the studied fishes.

2.2. *Triaenophorus nodulosus*

Tapeworms of the genus *Triaenophorus* are widespread parasites of freshwater fishes. They have a complex development cycle including definitive host and two aquatic intermediate hosts. Procercoids develop in the body cavity of their first intermediate host (Copepoda crustaceans). *Triaenophorus nodulosus* parasitize a wide range of second intermediate hosts and have been recorded from the fish belonging to 17 families. At plerocercoid stage, some species of this genus constitute a real danger to fishes, sometimes causing epizooties and even death of valuable commercial fish, mainly in fishery ponds (Kuperman, 1973).

In the Rybinsk reservoir and many other water bodies, the most common second intermediate host of *T. nodulosus* is the perch *Perca fluviatilis* (Kuperman, 1973). Plerocercoid *T. nodulosus* are extremely dangerous for young fish, particularly for the perch. They are localized in the liver where encapsulation occurs due to the proliferation of connective tissue around the parasite. Capsule formation is a protective response of the host organism to the parasite invasion (Kuperman, 1973). According to some reports, the host response directly (but disproportionately) depends on the invasion intensity. The less pronounced the parasite's specifity for the particular host species, the heavier pathologies arouse in the host organism, even at low intensity of invasion (Pronina & Pronin, 1988). The capsule that surrounds the parasite inside the host liver acts as a semipermeable membrane, owing to high activity of its wall and due to abundant capillary supply. On the one part, it provides favorable conditions for nutrition, growth and development of a parasite; on the other part, it safely protects the host tissues from its influence (Kuperman, 1973). The definitive host of *T. nodulosus* is the pike *Esox lucius* (L.) in whose intestine the parasite completes its development.

2.2.1. Triaenophorus nodulosus – Perch

The tests performed demonstrated that the glycogen content in the liver of the young perch harboring plerocercoid *T. nodulosus* declined from 30 ± 10 to 20 ± 1 mg/g. Besides, the infection was shown to cause a 42% increase in the wet weight of the liver and an equivalent (42%) decrease in liver glucose (Izvekova, 2001 b). These changes in carbohydrate metabolism suggest intensive carbohydrate utilization in response to infection.

Table 4. The activities of proteinases (PA) and glycosidases (GA) in the intestinal mucosa of perch juveniles infected with plerocercoid *T. nodulosus* (μmol/g/min)

Group of fish	PA		GA	
	pyloric caeca	intestine	pyloric caeca	intestine
Uninfected	6.78 ± 0.81*	13.0 3 ± 0.12	9.68 ± 2.51*	18.17 ± 4.72
Infected	4.25 ± 0.61*	14.03 ± 1.21	2.83±0.42*	10.32 ± 1.45

Mean ± SE indicated. Asterisks show significant differences between the test and control samples ($p < 0.05$).

Additionally, the proteinase and glycosidase activities were determined in the intestine and pyloric caeca of parasitized and unparasitized perch juveniles. In both groups, the activities of the studied enzymes in the pyloric caeca were significantly lower ($p < 0.05$) than those in the intestine (Table 4). The infection of young perch with plerocercoid *T. nodulosus* greatly affected the activities of digestive enzymes. In the infected specimens, both the activities of proteolytic and glycolytic enzymes were markedly reduced compared to uninfected controls (Table 4). Significant differences ($p < 0.05$) were noted for most comparisons. The sole exception was intestinal proteinase activity that hardly differed in the infected and uninfected fish.

The calculations showed the relative activity of the studied enzymes in pyloric caeca and intestine (A_{pc}/A_i) to decline in the infected fish (Table 5). It means that plerocercoids have greater influence on the enzyme activity in the pyloric caeca compared to that in the intestine. This could be explained by hepatopancreas malfunction in the parasitized individuals, which involves reductions in synthesis of proteolytic zymogens and amylase, and lowered intake of these substances into pyloric caeca. In addition, a two-fold decrease in the G/P ratio was observed in the paratisized fish, suggesting the greater impact of infection on the synthesis of glycosidases relative to proteinases (Table 5).

Table 5. Influence of plerocercoid *T. nodulosus* on the proportion between the glycosidase and protease activities in the young-of-the-year perch

Group of fish	G/P		A_{pc}/A_i	
	pyloric caeca	intestine	pyloric caeca	intestine
Uninfected	1.4	1.4	0.52	0.53
Infected	0.7	0.7	0.30	0.27

G/P ratio is glycosidase activity divided by the protease activity; A_{pc}/A_i ratio is enzyme activity in the pyloric caeca divided by the activity in the intestine. Mean ± SE indicated.

The negative influence of the plerocercoid *T. nodulosus* on the host organism similar to that revealed in our studies was noted also by other authors. For example, Kuperman (1973) examined a proportion between the liver weight and cumulative mass of parasites being indicative of parasite-induced injury to the perch organism. In the infected perch, this index varied from 3% to 60%. Besides, the perch harboring plerocercoids had lower red and higher white blood count. The presence of *T. nodulosus* in the liver of fishes (pike *Esox lucius*, Arctic cisco *Coregonus autumnalis*, and Arctic grayling *Thymallus arcticus*) caused local atrophy, fragmentation and necrosis of hepatic framework (Sidorov et al., 1989).

According to other reports, changes in the liver are feebly marked and associated mainly with damage and replacement of the parenchyma cells following the penetration of plerocercoids and capsule formation. In the presence of mature capsules around the plerocercoids, the hepatocytes display no histomorphologic injuries, and, obviously, no functional disorders, as the levels of RNA, protein, and glycogen in the infected liver are close to normal (Pronina & Pronin, 1988). In the opinion of these authors, the minor changes in the liver associated with plerocercoid encapsulation mainly account for the mechanical damage of hepatic tissues. This type of host tissue response to invasion testifies to a relative equilibrium in the considered host–parasite system and their mutual adaptation as well (Logachev & Pronin, 1975). At the same time, some decrease in body weight of the perch was noted following its infection with the plerocercoid *T. nodulosus*, as well as disease and mortality cases in the juveniles testifying to a higher pathogenicity of these helminths for younger fish (Pronina & Pronin, 1988).

It is also noteworthy that the degree of mutual adaptation of *T. nodulosus* and the perch may vary in different host populations, depending on the nature of helminth circulation in particular ecosystems. According to Pronina and

Pronin (1988), this fact provides an explanation for the inconsistent data on the parasite pathogenicity for the perch dwelling in different water bodies.

Judging by our results and some published data (Kuperman, 1973; Pronina & Pronin, 1988), the plerocercoid *T. nodulosus* cause substantial alterations in the perch liver, which may lead to host mortality at intensive invasion.

The effect of parasitic infection on the host glycogen content was also shown for other helminths. Thus, liver glycogen in carp *Cyprinus carpio* was found to depend on the intensity of its infection with *Bothriocephalus gowkongensis* decreasing proportionally to infection intensity (Davydov & Kurovskaya, 1991). Severe glycogen reduction occurred in the white muscles of the spotted porcupine fish *Diodon hystrix* following the infestation of their liver with plerocercoids of *Gymnorhynchus gigas* (Radhakrishnan et al., 1983). Glycogen depletion was also observed in the intestine of Philippine catfish *Clarius batrachus* infected with the common caryophillid parasite *Lytocestus indicus* (Rasheed & Simha, 1982).

2.2.2. Triaenophorus nodulosus – Pike

In case of the pike, being the definitive host for *T. nodulosus*, the glycogen concentrations were determined in the liver of parasite-free and infected specimens. The liver glycogen was shown to decrease from 0.55 ± 0.04 to 0.49 ± 0.04 mg/g following infection, however the difference was insignificant. Field observations have demonstrated that even heavy infestation of the pike with *T. nodulosus* caused no death or notable emaciation of fish. A fairly low virulence of the cestode relative to the pike may evidence the balance in this old-established and steady host–parasite system (Kuperman, 1973). Therefore changes in the activity of digestive enzymes in the pike harboring adult *T. nodulosus* reported in our recent studies (Izvekova & Solovyev, 2012), unlike those following the plerocercoid infection, are probably not associated with hepatopancreas dysfunctions.

Thus, parasitization by plerocercoid *T. nodulosus* tends to depress the liver glycogen in second intermediate host, while adult worms living in the intestine of their definitive host have no marked effect on the liver glycogen reserves.

CONCLUSION

The following major conclusions can be drawn from our results and their comparison with those obtained by other authors. The parasitization of fishes by helminthes influences the glycogen content in host tissues. The muscle glycogen tends to increase following the experimental and natural infection with trematode metacercaria in a number of cyprinid species. Severe reductions in spontaneous motor activity of fish maintained in the presence of trematode-infected mollusks can be responsible for the glycogen rise. In turn, a decrease in fish swimming activity may account for the muscle injuries caused by the trematode metacercaria.

In contrast, fish harboring the plerocercoid cestodes have lower muscle and liver glycogen compared to uninfected controls. The glycogen depletion in the hepatopancreas testifies to a dysfunction of this organ in parasitized fish and, as a result, leads to a decrease in the activity of the major digestive enzymes (proteinases and glycosidases) in the intestine. The decrease in the activity of digestive enzymes affects the efficiency of food assimilation in fish. On the contrary, the presence of adult worms in the fish intestine hardly influences the liver glycogen.

Thus, the infestation with trematodes and cestodes appears to have a different effect on the tissue glycogen levels in fish and so might differently influence their physiological state. However, it should be born in mind that helminth-induced alterations in carbohydrate metabolism have been studied in rather few taxa of fishes and their parasitic organisms. Besides, the possible effect of invasion intensity along with a dynamic host response to parasitic infection must be taken into account when comparing the glycogen variations caused by these two phylogenetically distant groups of helminths.

Acknowledgments

These studies were supported in part by the Russian Foundation for Basic Research, project no. 10-04-00204-a. The authors are grateful to Valentin A. Nepomnyashchikh (Institute for Biology of Inland Waters RAS) and Alexander O. Mostryukov (Institute of Physics of the Earth RAS) who developed the software for the analysis of video records.

References

Amat, F., Gozalbo, J. C., Navarro, J. C., Hontoria, F., Varó, I. (1991). Some aspects of *Artemia* biology affected by cestode parasitism. *Hydrobiologia, 212*, 29–44.

Anson, M. (1938). The estimation of pepsin, tripsin, papain and cathepsin with hemoglobin. *J. Gen. Physiol., 22(1),* 79–83.

Arakelova, E. S., Chebotareva, M. A., Zabelinskii, S. A. (2003). On adaptative changes of rates of oxygen consumption and lipid metabolism in *Littorina saxatilis* at parasitic invasions. *J. Evol. Biochem. Physiol., 39(5),* 519–528.

Arakelova, K. S., Chebotareva, M. A., Zabelinskii, S. A. (2004). Physiology and lipid metabolism of *Littorina saxatilis* infected with trematodes. *Dis. Aquat. Org., 60,* 223–231.

Arme, C., & Owen, R. W. (1968). Occurrence and pathology of *Ligula intestinalis* (L.) infections in British fishes. *J. Parasitol., 54,* 272–280.

Arme, C., & Owen, R. W. (1970). Observations on a tissue response within the body cavity of fish infected with the plerocercoid of *Ligula intestinalis* (L.) (Cestoda). *J. Fish Biol., 2,* 35–37.

Baldwin, N. L., Millemann, R. E., Knapp, S. E. (1967). "Salmon poisoning" disease. III. Effect of experimental *Nanophyetus salmonicola* infection on the fish host. *J. Parasitol., 53(3),* 556–564.

Barber, I. (2003). The role of parasites in fish-bird interactions: a behavioural ecological perspective. In I. G. Cowx (Ed). Interactions between fish and birds: Implications for management (pp. 221–243). Oxford: Blackwell Science.

Barber, I., & Wright, H. A. (2006). Effects of parasites on fish behaviour: Interactions with host physiology. In K. Sloman, R. W. Wilson, & S. Balshine (Eds.). Behaviour and physiology of fish. Vol. 24 (pp. 109–149). San Diego: Elsevier/Academic Press.

Barber, I., Hoare, D., Krause, J. (2000). Effects of parasites on fish behavior: review and evolutionary perspective. *Rev. Fish Biol. Fish., 10,* 131–165.

Barrett, K. W. (1978). Studies on beta-oxidation in the plerocercoids of *Ligula intestinalis* (Cestoda: Pseudophyllidea). *Z. Parasitenk., 57(3),* 243–246.

Baruš, V., Šebela, M., Prokeš, M. (1997). On the possible free-living phase of the *Ligula intestinalis* plerocercoids (Ligulidae). *Helminthologia, 34(3),* 173–174.

Biserova, L.I. (1990). Occurrence and distribution of *Lithoglyphus naticoides* (Gastropoda, Lithoglyphidae) in the Volga River delta. *Gidrobiol. Zh., 26(2),* 98–100.

Blake, R. W., Kwok, P. Y. L., Chan, K. H. S. (2006). Effects of two parasites, *Schistocephalus solidus* (Cestoda) and *Bunodera* spp. (Trematoda), on the escape fast-start performance of three-spined sticklebacks. *J. Fish Biol., 69,* 1345–1355.

Brassard, P., Rau, M. E., Curtis, M. A. (1982) Parasite-induced susceptibility to predation in diplostomoiasis. *Parasitology, 85,* 495–501.

Butler, J. A., & Millemann, R. E. (1971). Effect of the "salmon poisoning" trematode, *Nanophyetus salmonicola*, on the swimming ability of juvenile salmonid fishes. *J. Parasitol. 57(4),* 860–865.

Davydov, O. N., & Kurovskaya, L. Ya., (1991). Host–parasite relationships in fish cestodosis. Kiev: Naukova Dumka.

Dubinina, M. N. (1980). Tapeworms (Cestoda, Ligulidae) of the fauna of the USSR. New Delhi: Amerind Publishing Co. Ltd

Coban, M.Z., & Sen D. (2011). Examination of liver and muscle glycogen and blood glucose levels of *Capoeta umbla* (Heckel, 1843) living in Hazar Lake and Keban Dam Lake (Elazig, Turkey). *Afr. J. Biotechnol., 10(50),* 10271–10279.

Hochachka, P. W. (1985). Fuels and pathways as designed systems for support of muscle work. *J. Exp. Biol., 115,* 149–164.

Izvekov, E. I., Izvekova, G. I., Tyutin, A. V. (2012). Glycogen content and motor activity in cyprinid fishes infected with metacercariae of the trematode *Apophallus muehlingi*. Modern problems of general parasitology. The Materials of the International Conference (30 October – 1 November, 2012, Moscow) (pp. 142–145). Moscow: Parasitological Center of the A. N. Severtsov Institute of Ecology and Evolution RAS.

Izvekova, G. I. (2001 a). *Ligula intestinalis* (Cestoda, Pseudophyllidea): Some aspects of plerocercoids metabolism. *Biology of Inland Water, 2,* 101–106.

Izvekova, G. I. (2001 b). Physiological specificity of relationships between *Triaenophorus nodulosus* (Cestoda) of and its fish host. *Parazitologiya, 35(1),* 60–68.

Izvekova, G. I., & Kuz'mina, V. V. (1996). The influence of the plerocercoid *Ligula intestinalis* infection on the activity of digestive enzymes and glycogen content in tissues of the bream. *Parazitologiya, 30(1),* 45–51.

Izvekova, G. I., & Solovyev, M. M. (2012). The activity of digestive enzymes of the pike *Esox lucius* L. Infected with the Cestode *Triaenophorus nodulosus* (Pallas). *Inland Water Biol., 5(1),* 113–118.

Izvekova, G. I., & Tyutin, A. V. (2011 a). Infection and parasite–host relation peculiarities in the system *Ligula intestinalis* – sabrefish (*Pelecus cultratus*) of in the Rybinsk reservoir. *Povolzhskiy J. Ecol., 2,* 137–145.

Izvekova, G. I., & Tyutin, A. V. (2011 b). Occurrence of partenites in mollusks and the influence that metacercaria of *Apophallus muehlingi* (Jagerskiold, 1898) and *Posthodiplostomum cuticola* (Nordmann, 1832)

has on some biochemical parameters in fish. *Inland Water Biol., 4(3),* 367–372.

Izumova, N. A. (1987). Parasitic fauna of reservoir fishes of the USSR and its evolution. New Dehli: Amerind Publishing Co. Ltd.

Johnston, I. A., & Goldspink, G. (1973 a). A study of the swimming performance of the crucian carp *Carassius carassius* (L.) in relation to the effects of exercise and recovery on biochemical changes in the myotomal muscles and liver. *J. Fish Biol., 5,* 249–260.

Johnston, I. A., & Goldspink G. (1973 b). Quantitative studies of muscle glycogen utilization during sustained swimming in crucian carp (*Carassius carassius* L.). *J. Exp. Biol., 59,* 607–615.

Johnston, I.A., Davison, W., Goldspink G. (1977). Energy metabolism of carp swimming muscles. *J. Comp. Physiol. B, 114,* 203–216.

Klein, W. D., Olsen, O. W., Bowden, D. C. (1969). Effects of intestinal fluke, *Crepidostomum farionis*, on rainbow trout, *Salmo gairdnerii. Trans. Am. Fish. Soc., 98(1),* 1–6.

Kosareva, N. A. (1961). Metabolic disturbances in cyprinid fishes as a result of their invasion by *Ligula* and *Digramma*. *Dokl. Akad. Nauk SSSR, 139,* 510–512.

Kublickiene, O., & Goldbergiere, M. (1966). Hystochemical and biochemical studies on metabolism in experimental fascioliasis. *Acta Parasitol. Litu., 6,* 98–99.

Kuperman, B. I. (1973). Tapeworms of the genus *Triaenorhorus* — fish parasites. Leningrad: Nauka.

Kuperman, B. I., Zhokhov, A. Ye., Izvekova, G. I., Talikina, M. G. (1997). Dynamics of bream invasion by *Ligula intestinalis* in the Volga reservoirs and their parasite-host relations. *Biology of Inland Water, 2,* 41–49.

Kurovskaya, L. Ya. (2002). Effects of mixed invasion on the protein content in organs of carp yearlings reared in warm water. *Hydrobiol. J., 38(1),* 1–9.

Kurovskaya, L. Ya. (1993). The effect of lower cestodes of the order *Pseudophyllidea* on the viability of two-year-old white amur. *Parazitologiya, 27(1),* 59–68.

Lemly, A. D., & Esch, G. W. (1984). Effects of the trematode *Uvulifer ambloplitis* on juvenile bluegill sunfish, *Lepomis macrochirus*: ecological implications. *J. Parasitol., 70,* 475–492.

Li, J., Liao, X., Yang, H. (2000). Molecular characterization of a parasitic tapeworm (*Ligula*) based on DNA sequences from formalin-fixed specimens. *Biochem. Genet., 38(9/10),* 309–322.

Liew, H. J., Sinha, A. K., Mauro, N., Diricx, M., Blust, R., De Boeck, G. (2012). Fasting goldfish, *Carassius auratus*, and common carp, *Cyprinus carpio*, use different metabolic strategies when swimming. *Comp. Biochem. Physiol. A, 163(3–4),* 327–335.

Logachev, E. D., & Pronin, S. V. (1975). The microstructural characteristic of the interactions in the host–parasite system at simultaneous parasitism of fishes by nematode larvae and cestodes. In B.N. Mazurovich (Ed.) Parasites and parasitoses of animals and humans (pp. 42–47). Kiev: Naukova dumka.

Loot, G., Brosse, S., Lek S., Guegan, J.-F. (2001). Behaviour of roach (*Rutilus rutilus* L.) altered by *Ligula intestinalis* (Cestoda: Pseudophyllidea): field demonstration. *Freshwat. Biol., 46(9),* 1219–1227.

Lowe, T. E., & Wells, R. M. G. (1997). Exercise challenge in Antarctic fishes: do hematology and muscle metabolite levels limit swimming performance? *Polar. Biol., 17,* 211–218.

Luo, H. Y., Nie, P., Yao, W. J., Wang, G.T., Gao, Q. (2003). Is the genus *Digramma* synonymous to the genus *Ligula* (Cestoda: Pseudophyllidea)? *Parasitol. Res., 89,* 419–421.

Lutterschmidt, W. I., Schaefer, J. F., Fiorillo, R. A. (2007). The ecological significance of helminth endoparasites on the physiological performance of two sympatric fishes. *Comp. Parasitol., 74(2),* 194–203.

Markov, G. S., Kubantzev, B. S., Markova, E. K., Vakhmenina, V. I., Kosareva, N. A. (1978). The effect of invasion by *Ligula intestinalis* on morphophysiological features in *Abramis brama*. *Russ. J. Ecol., 2,* 32–36.

Matskási, I., & Juhász, S. (1977). *Ligula intestinalis* (L., 1758): Investigation on plerocercoids and adults for protease and protease inhibitor activities. *Parasitol. Hung., 10,* 51–59.

Matskási, I., & Németh, I. (1979). *Ligula intestinalis* (Cestoda: Pseudophyllidae): studies on the properties of proteolytic and protease inhibitor activities of plerocercoid larvae. *Int. J. Parasitol., 9(3),* 221–227.

McManus, D. P. (1975). Tricarboxylic acid cycle enzymes in the plerocercoid of *Ligula intestinalis* (Cestoda: Pseudophyllidea). *Z. Parasitenk., 45(4),* 319–322.

McManus, D. P., & Sterry, P. R. (1982). *Ligula intestinalis*: intermediary carbohydrate metabolism in plerocercoids and adults. *Z. Parasitenk., 67,* 73–85.

Montgomery, R. (1957). Determination of glycogen. *Arch. Biochem. Biophys., 67,* 378–386.

Moyes C. D., & West, T.G. Exercise metabolism of fish. In P. W. Hochachka, & T. P. Mommsen (Eds.). Biochemistry and molecular biology of fishes. Vol. 4. Metabolic biochemistry (pp. 367–392). Elsevier Science. B. V.

Odening, K. (1970). Der Entwicklungszyklus von *Apophallus muehlingi* (Trematoda: Opisthorchiida: Heterophyidae) in Berlin. *Z. Parasitenk., 33,* 194–210.

Ottolenghi, C., Puviani, A. C., Brighenti, L. (1981). Glycogen in liver and other organs of catfish (*Ictalurus melas*): Seasonal changes and fasting effects. *Comp. Biochem. Physiol. A, 68(3),* 313–321.

Parkhouse, W.S., Dobson, G.P., Hochachka., P.W. (1988 a). Control of glycogenolysis in rainbow trout muscle during exercise. *Can. J. Zool., 66,* 345–351.

Parkhouse, W.S., Dobson, G.P., Hochachka, P.W. (1988 b). Organization of energy provision in rainbow trout during exercise. *Am. J. Physiol., 254(2),* 302–309.

Pinheiro, J., Maldonado, A. J., Lanfredi, R. M. (2009). Physiological changes in *Lymnaea columella* (Say, 1817) (Mollusca, Gastropoda) in response to *Echinostoma paraensei* Lie and Basch, 1967 (Trematoda: Echinostomatidae) infection. *Parasitol. Res., 106,* 55–59.

Plaistow, S. J., Troussard, J.-P., Cézilly, F. (2001).The effect of the acanthocephalan parasite *Pomphorhynchus laevis* on the lipid and glycogen content of its intermediate host *Gammarus pulex. Int. J. Parasitol., 31,* 346–351.

Plisetskaya, E. M. (1975). Hormonal regulation of carbohydrate metabolism in lower vertebrates. Leningrad: Nauka.

Plisetskaya, E. M., & Kuzmina, V. V. (1972). The glycogen content in the organs of Cyclostomata and Pisces. *J. Ichthyol., 12(2),* 335–343.

Ponton, F., Biron, D. G., Joly, C., Helluy, S., Duneau, D., Thomas, F. (2005). Ecology of populations parasitically modified: a case study from a gammarid (*Gammarus insensibilis*)-trematode (*Microphallus papillorobustus*) system. *Mar. Ecol. Progr., 299,* 205–215.

Pronina, S. V., & Pronin, N. M. (1988). Relationships in worm–fish systems (tissue, organ, and whole body levels). Moscow: Nauka.

Radhakrishnan, S., Nair, N. B., Balasubramanian, N. K. (1983). Gymnorhynchus gigas plerocercoid (Cestoda: Gymnorhynchidae) infection of the liver of *Diodon hystrix* (Pisces: Diodontidae) III. Biochemical composition of infected fish. *Acta Ichthyol. Piscat., XIII (2),* 141–147.

Rasheed, U., & Simha, S. S. (1982). Quantitative and qualitative studies of glycogen, proteins, lipids and phospholipids of *Clarius batrachus* with relation to the parasite *Lytocestus indicus* Moghe, 1931. *J. Anim. Morphol. Physiol., 29(1–2),* 232–235.

Reshetnikov, Yu. S., (Ed.). (2010). Fishes in the reserved areas of Russia. Vol. 1. Freshwater fishes. Moscow: KMK (Association of Scientific Publications).

Richards, K. S., & Arme, C. (1981). The effect of the plerocercoid larva of the pseudophyllidean cestode *Ligula intestinalis* on the musculature of bream (*Abramis brama*). *Z. Parasitenk., 65(2),* 207–215.

Rosen, V. B. (1984). Fundamentals of endocrinology. Moscow: Vysshaya Shkola.

Santos, E. G. N., Cunha, R. A., Santos, C. P. (2011). Behavioral responses of *Poecilia vivipara* (Osteichthyies: Cyprinodontiformes) to experimental infections of *Acanthocollaritrema umbilicatum* (Digenea: Cryptogonimidae). *Exp. Parasitol., 127,* 522–526.

Scarabello, M., Wood, C. M., Heigenhauseg, R. J. F. (1991). Glycogen depletion in juvenile rainbow trout as an experimental test of the oxygen debt hypothesis. *Can. J. Zool., 69,* 2562–2568.

Shchepkina, A. M. (1981). The effect of trematode metacercariae *Cryptocotyle concavum* on the lipid composition of goby. *Parazitologiya, 15(2),* 185–187.

Shields, B. A., Groves, K. L., Rombaugh, C., Bellmore, R. (2002). Ligulosis associated with mortality in largescale suckers. *J. Fish Biol., 61,* 448–455.

Sidorov, V. S, Vysotskaya, R. U., Smirnov, L. P., Guryanov, S. D. (1989). Comparative biochemistry of fish helminths. Leningrad: Nauka.

Simões, S. B., das Neves, R. F., Santos, C. P. (2008). Life history of *Acanthocollaritrema umbilicatum* Travassos, Freitas and Bührnheim, 1965 (Digenea: Cryptogonimidae). *Parasitol. Res., 103(3),* 523–528.

Smyth, J. D., & McManus, D. P. (1989). The physiology and biochemisty of cestodes, Cambridge: Cambridge University Press.

Sogandares-Bernal, F., Hietala, H. J., Gunst, R. F. (1979). Metacercariae of *Ornithodiplostomum ptychocheilus* (Faust, 1917) Dubois, 1936 encysted in the brains and viscera of red-sided shiners from the Clark-fork and Bitterroot rivers of Montana: An analysis of the infected hosts. *J. Parasitol., 65,* 616–623.

Sorvachev, K. F. (1982). Fundamentals of biochemistry of fish nutrition. Moscow: Legkaya i Pishchevaya Promyshlennost.

Soutter, A. M., Walkey, M., Arme, C. (1980). Amino acids in the plerocercoid of *Ligula* intestinalis (Cestoda: Pseudophyllidea) and its fish host, *Rutilus rutilus*. *Z. Parasitenk., 63(2),* 151–158.

Stadnichenko, A. P. (1978). Changes in the content of sulfhydryl groups in the hemolymph of freshwater gastropods invaded with parthenitic species and trematode larvae. *Parazitologiya, 12(1),* 79–82.

Sweeting, R. A. (1977). Studies on *Ligula intestinalis*. Some aspects of the pathology in the second intermediate host. *J. Fish Biol., 10,* 43–50.

Symons, L. E. A., & Jones, W. O. (1974). Basal metabolic rate and lipid and liver glycogen in mice infected with the nematode *Nematospiroides dubius*. *Int. J. Parasitol., 4,* 301–305.

Taylor, M., & Hoole, D. (1989). *Ligula intestinalis* (L.) (Cestoda: Pseudophyllidea): plerocercoidinduced changes in the spleen and pronephrous of roach, *Rutilus rutilus* (L.) and gudgeon, *Gobio gobio* (L.). *J. Fish Biol., 34(4),* 583–596.

Thillart van den, G., & Smit, H. (1984). Carbohydrate metabolism of goldfish (*Carassius auratus* L.): Effects of long-term hypoxia-acclimation on enzyme patterns of red muscle, white muscle and liver. *J. Comp. Physiol. B, 154,* 477–486.

Thillart van den, G., & Raaij van, M. (1995). Endogenous fuels; non-invasive versus invasive approaches. In P. W. Hochachka, & T. P. Mommsen (Eds.). Biochemistry and molecular biology of fishes. Vol. 4. Metabolic biochemistry (pp. 33–63). Elsevier Science. B. V.

Thompson, S. N. (1983). Biochemical and physiological effects of metazoan endoparasites and on their host species. *Comp. Biochem. Physiol. B, 74,* 183–211.

Thompson, S. N. (1997). Physiology and biochemistry of snail-larval trematode relationships. In B. Fried, & T. K. Graczyk (Eds.) Advances in trematode biology (pp. 149–195). Boca Raton - New York: CRC Press.

Trenzado, C. E., Morales, A. E., de la Higuera, M. (2008). Physiological changes in rainbow trout held under crowded conditions and fed diets with different levels of vitamins E and C of and highly unsaturated fatty acids (HUFA*). Aquaculture, 277(3–4),* 293–302.

Tsuboi, T., Hirai, K., Torii, M., Nishida, H. (1991). Decrease of liver glycogen content in golden hamsters infected with plerocercoids of *Spirometra erinacei*. *Parasitol. Res., 4,* 320–324.

Tyutin, A. V., & Izvekova, G. I. (2013). Infection of mollusks and fish by the trematode *Apophallus muehlingi* (Jagerskiold, 1898) and its interrelations with intermediate hosts. *Inland Water Biol., 6(1),* 52–56.

Ugolev, A. M, Iezuitova, N. N., Masevich, Ts. G., Nadirova, T. Ya., Timofeeva, N. M. (1969). Study of the digestive apparatus in humans (A review of current methods). Leningrad: Nauka.

Vanacker, M., Masson, G., Beisel, J-N. (2012). Host switch and infestation by *Ligula intestinalis* L. in a silver bream (*Blicca bjoerkna* L.) population. *Parasitology, 139,* 406–417.

Weber, J.-M. (1992). Pathways for oxidative fuel provision to working muscles: Ecological consequences of maximal supply limitations. *Experientia, 48,* 557–564.

Weber, J.-M., & Haman, F. (1996). Pathways for metabolic fuels and oxygen in high performance fish. *Comp. Biochem. Physiol. A, 113(1),* 33–38.

Weber, J.-M., Roberts, T. J., Vock, R., Weibel, E. R., Taylor, C. R. (1996). Design of the oxygen and substrate pathways. III. Partitioning energy provision from carbohydrates. *J. Exp. Biol., 199(8),* 1659–1666.

In: Glycogen ISBN: 978-1-62948-395-5
Editors: P.L. Weiss and B.D. Faulkner

Chapter 4

GLYCOGEN METABOLISM ENZYMES AS MOLECULAR TARGETS FOR DRUG DEVELOPMENT

Vassiliki T. Skamnaki, Anastassia L. Kantsadi, Demetra S. M. Chatzileontiadou, Georgios Stravodimos and Demetres D. Leonidas*

Department of Biochemistry and Biotechnology,
University of Thessaly, Greece

ABSTRACT

Glycogen, the cellular energy store, is associated with a set of enzymes controlling its metabolism in a complex and dynamic way. Although glycogen metabolism has been thorough investigated, the regulatory mechanisms governing this process are not fully understood to date. New ideas have emerged in an effort to elucidate the complications of regulation of glycogen metabolism, from altered phosphorylation status and enzyme translocation to the role of intracellular compartmentalization and heterogeneity of glycogen stores as a regulator of metabolic pathways, and signalling transduction and lysosomal

* To whom correspondence should be addressed: Assoc. Prof. Demetres D. Leonidas, Department of Biochemistry and Biotechnology, University of Thessaly, 26 Ploutonos Str. 41221 Larissa, Greece. Phone: +30 2410 565278; Fax: +30 2410 565290; e-mail: ddleonidas@bio.uth.gr.

disposal of glycogen. The correlation of glycogen metabolism to many metabolic disorders has targeted many of the enzymes involved, for the development of drugs for a number of metabolic diseases, predominantly diabetes mellitus but also cancer. The notion of rational drug design emerged as the most appropriate modern interdisciplinary approach. This approach involves the close collaboration of diverse fields from bioinformatics and organic synthesis to structural biology (X-ray crystallography, NMR and EM) and physiology. In this review we will attempt to highlight major aspects and recent progress in the research for new drugs targeting enzymes playing a dominant role in the degradation rather than the synthetic glycogen pathway, such as glycogen phopshorylase and glycogen phosphorylase kinase.

INTRODUCTION

Glycogen metabolism is a relatively simple biochemical process within the cells for the production of an oxidizable energy source, glucose. In organisms, glycogen metabolism is the main mechanism for glucose homeostasis. However, the regulation of glycogenolysis and synthesis of glycogen is a rather complex mechanism which involves the interplay of several enzymes and a phosphorylation/dephosphorylation cascade. Glycogen synthesis and degradation is determined by glycogen synthase (GS) and glycogen phosphorylase (GP), respectively. The activity of GS is inhibited by the presence of GPa (a phosphorylated form of GP which is fully active) since GPa is a potent inhibitor of glycogen synthase phosphatase which catalyses the dephosphorylation and activation of GS. This way, the phosphorylation state of GP determines not only the rate of glycogenolysis but also the rate of glycogen synthesis [1]. Thus, phosphorylase kinase (PhK) the enzyme that catalyses the phosphorylation of GP has a pivotal role in the mobilization of glycogen storage and glucose homeostasis.

GP (EC 2.4.1.1) and glycogen phosphorylase kinase (PhK) have a pivotal role in glycogen breakdown and hence in glucose homeostasis. GP catalyses the phospholytic breakdown of the α-1,4 glycosidic bonds in glycogen to produce glucose-1-phosphate (Glc-1-P). It starts from the non-reducing end of glycogen and continues until it reaches four glucose residues away from α-1,6 branch. GP has a cofactor, pyridoxal 5-phosphate (PLP, derived from vitamin B_6) linked via a shiff base to Lys680, at the active site [2]. In response to adrenaline or glucagon PhK (EC 2.7.1.38) catalyzes the Ca^{2+} dependent phosphorylation of a serine residue (Ser14) in GP and thus transforming GPb (low activity, low substrate affinity) to the GPa (high activity, high substrate affinity) [2]. PhK is activated also by phosphorylation from protein kinase A which is activated by cyclic AMP. The reverse reaction of dephosphorylation that inactivates GP is catalyzed by protein phosphatase 1 (PP1) [3], an enzyme that is regulated in response to insulin. The conversion of GPa to GPb relieves the allosteric inhibition that GPa exerts on the glycogen-associated PP1, which converts glycogen synthase D to the I form, thus allowing the phosphatase to stimulate the synthesis of glycogen [2].

The correlation of glycogen metabolism to many metabolic disorders has targeted GP and PhK, for the development of drugs, for a number of metabolic diseases, predominantly diabetes mellitus but also cancer. Pharmacological modulation of the glycogen metabolism and the consequences of inhibiting glycogenolysis are a continuing challenge for physiology, biochemistry and pharmacology [4]. The number of drug related patents filled by pharmaceutical and biotechnology companies for these two enzymes is steadily increasing in the last years [5].

Glycogen Phosphorylase Overview

GP is an allosteric enzyme, which exists in three isoforms liver, muscle and brain [6]. GP is a regulatory enzyme that catalyses the first step in glycogen degradation to yield Glc-1-P. In the muscle, Glc-1P is utilized via glycogenolysis to generate metabolic energy. In the liver, Glc-1-P, is mostly converted by phosphoglucomutase and glucose-6-phosphatase to glucose, which is released for the benefit of other tissues [6], especially the central nervous system that relies on glucose as its major source of fuel. Following the MWC allosteric model [7] GP exists in two conformational states T (low activity and affinity for substrates) and R (high activity and affinity for substrates). The equilibrium between the two states can be altered by the presence of allosteric modulators that stabilize one form or the other, regulating thus its activity. Further to this, GP activity is further regulated by the phosphorylation of Ser14 by PhK, which transforms the inactive GPb to active GPa [8, 9].

GP is a homodimer, composed by two identical subunits of MW 97.5 kD each, that also contain the cofactor PLP (Figure 1). The structures of T and R state of GPa and GPb have been determined by X-ray crystallography [8, 10-12].

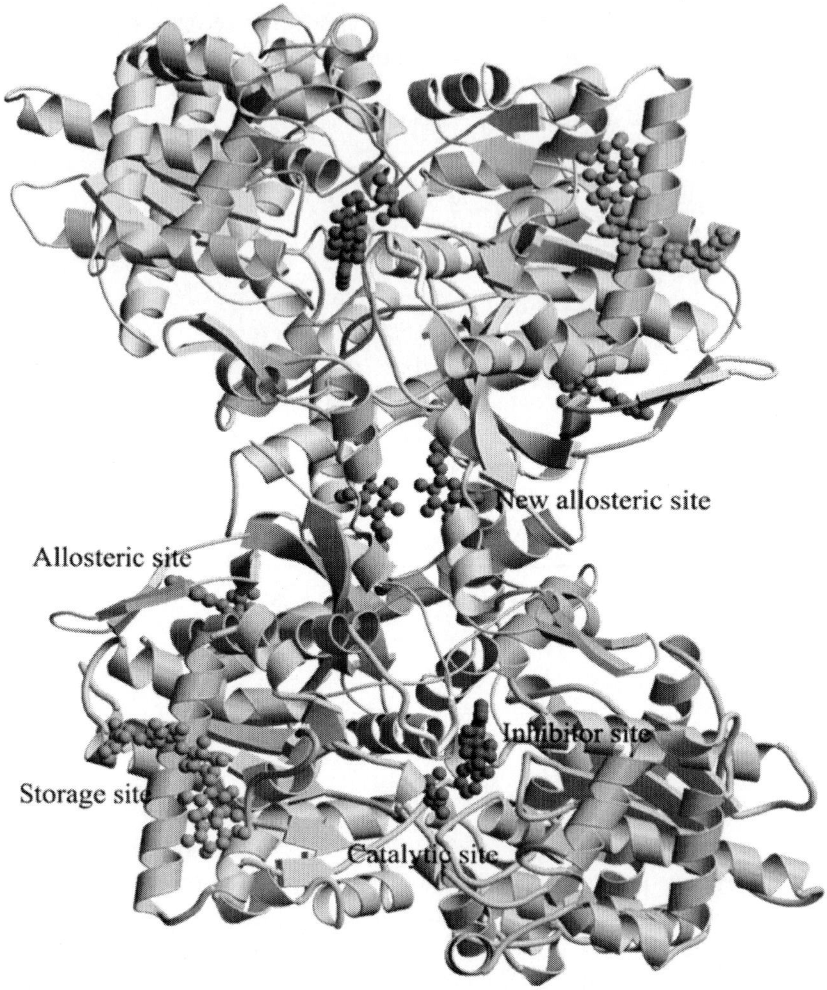

Figure 1. A schematic diagram of the muscle GPb dimeric molecule. The positions for the catalytic, allosteric, inhibitor, and the new allosteric site positions are shown by the binding of glucose, AMP, caffeine, maltoheptaose and a chloroindole-carboxamide inhibitor, respectively.

In the muscle enzyme, conversion from the T to R state by phosphorylation or AMP, induces a major quaternary structural change which results to a rotation of each subunit by 5° about an axis at the dimer interface [8]. In the T-state enzyme there is no access to the catalytic site from the surface since this is blocked by a loop termed 280s (residues 282 to 286). On transition from T-state to R-state, the 280s loop becomes disordered and displaced, thus opening a channel that allows access for glycogen to reach the active site from the surface [2]. Kinetic and X-ray crystallography studies have identified several ligand binding sites in GP. These include the Ser14-phosphate recognition site, the catalytic site that binds substrates Glc-1-P and glycogen but also the physiological inhibitor glucose, the inhibitor site, which binds caffeine and related compounds, the glycogen storage site, the allosteric site that binds the activator AMP, and an allosteric indole-binding site at the dimer interface [13, 14]. The overall architecture of the native T-state muscle GPb with the various binding sites is shown in Figure 1, with the two subunits of the functionally active dimer related by a 2-fold symmetry axis.

Glycogen Phosphorylase Active Site Inhibitors

A plethora of high resolution GP structural data (190 structures have been deposited in the Protein Data Bank thus far) has trigger rational structure based design efforts for the discovery of novel inhibitors. These efforts are currently summarized in several excellent reviews [2, 13-16] that have described the progress made in the past. The efficacy of such inhibitors on blood glucose control and hepatic glycogen balance has been confirmed from animal studies and *ex vivo* cell biology experiments [1, 4, 17-19] validating GP as an important target for the design of new hypoglycemic agents. Here, we will highlight the recent advances in the quest for a potent inhibitor with high specificity for the GP active site, which are not covered by the aforementioned reviews. This quest has now shifted towards molecules that bare drug-like properties in an attempt to increase the efficacy of these inhibitors at cellular level.

It has to be noted that whereas the physiological target for inhibitors is the human liver GP, the muscle isoform is widely used as a good surrogate, given that the two enzymes are 79% identical with 100% identity in amino-acid sequence and structural architecture at the catalytic site [20, 21]. Hence, most of the studies have been performed on the muscle enzyme but the conclusions drawn can be applied in human liver glycogen phosphorylase. Because of the homology between the three isoforms of GP it may not be possible to achieve selectivity for the liver isoform. However, since the liver is the major site of exposure to drugs administered orally, it should be possible to achieve an appropriate drug dosage that targets predominantly hepatic GP [1].

Table 1. The chemical structures of the most potent inhibitors of the GP catalytic site discovered to date, together with the K_i values for each inhibitor

	Chemical Structure	K_i (μM)
1	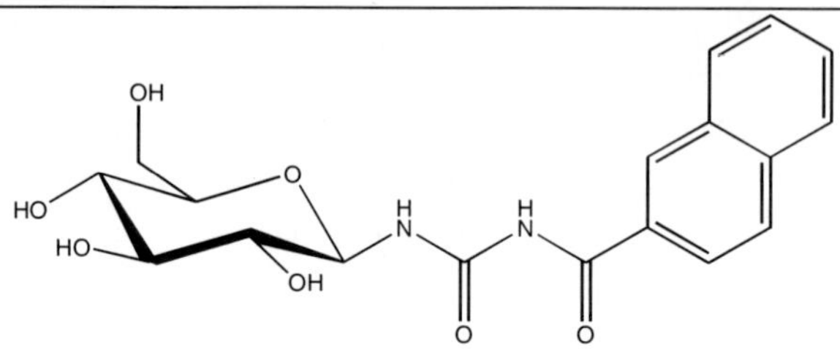	0.35 [16]
2	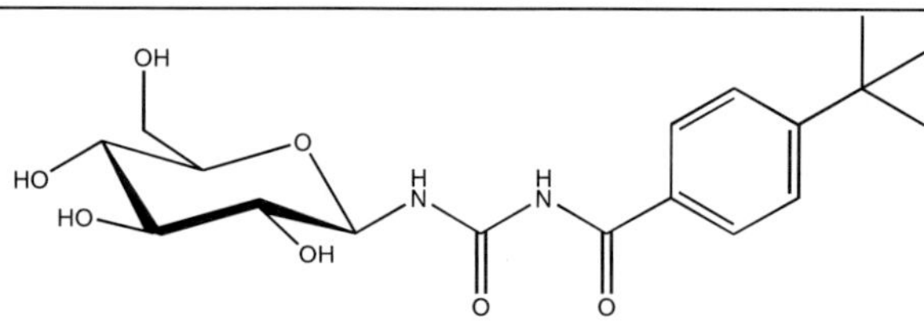	0.7 [14]
3	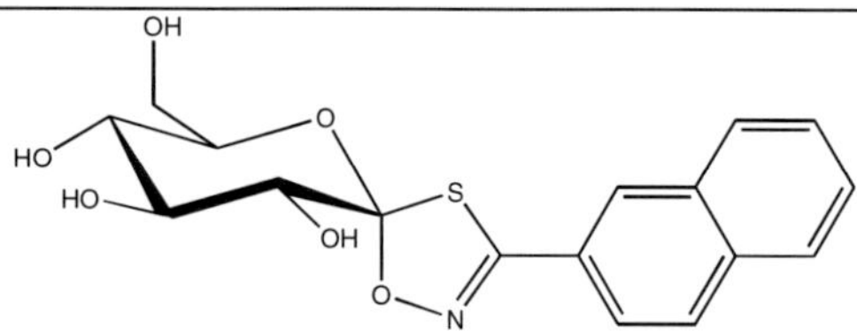	0.16 [16]
4	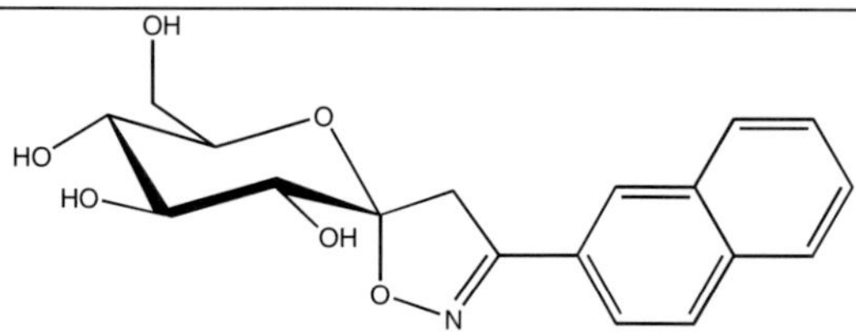	0.63 [89]
5	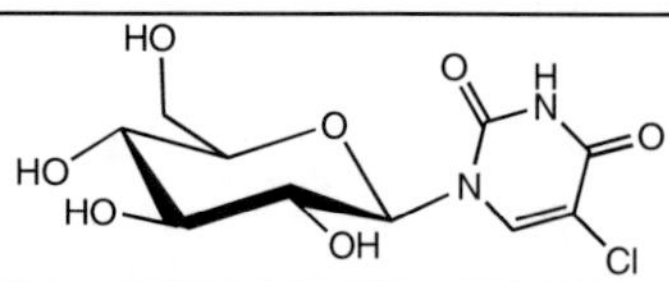	1.02 [26]
6	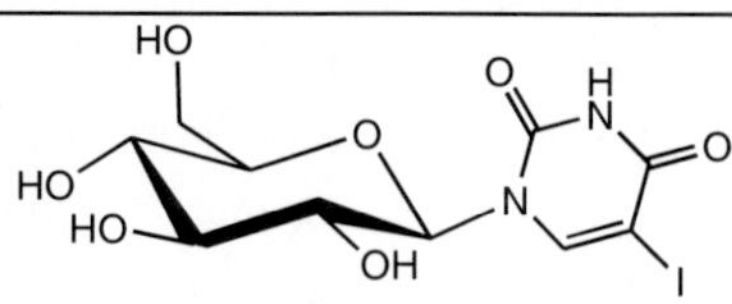	1.94 [26]

	Chemical Structure	Ki (μM)
7	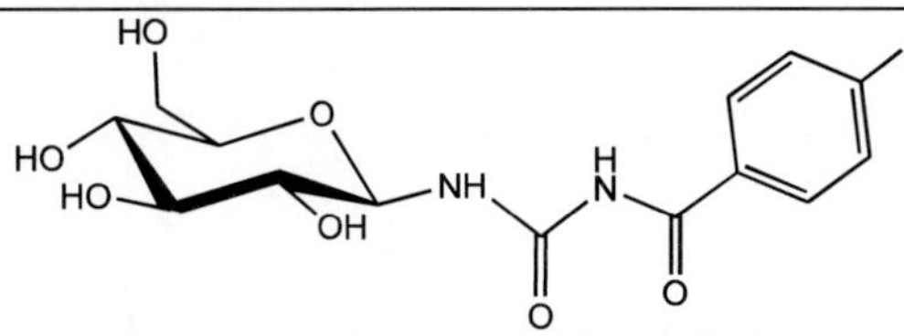	2.3 [27]
8	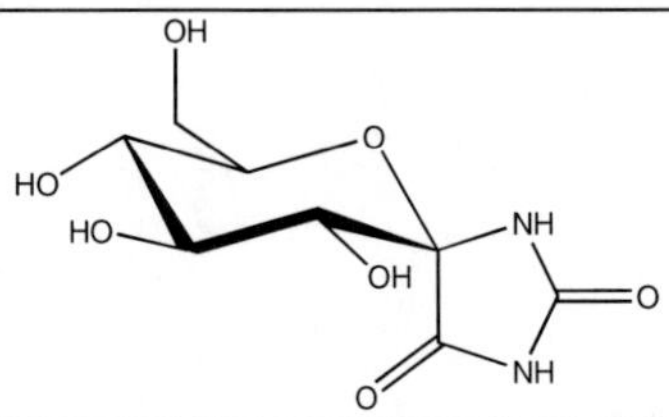	3.1 [27]
9	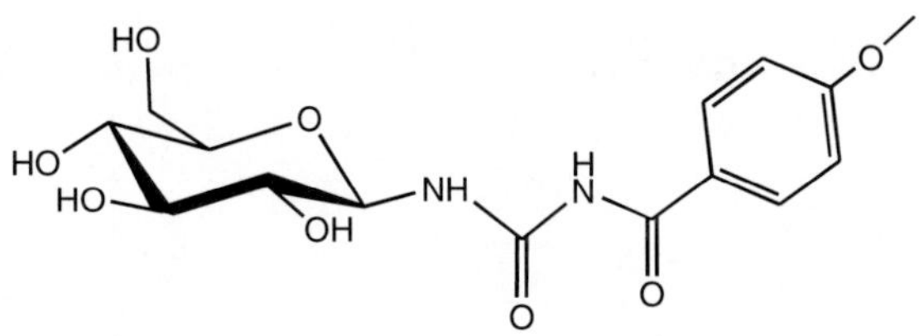	3.2 [27]
10	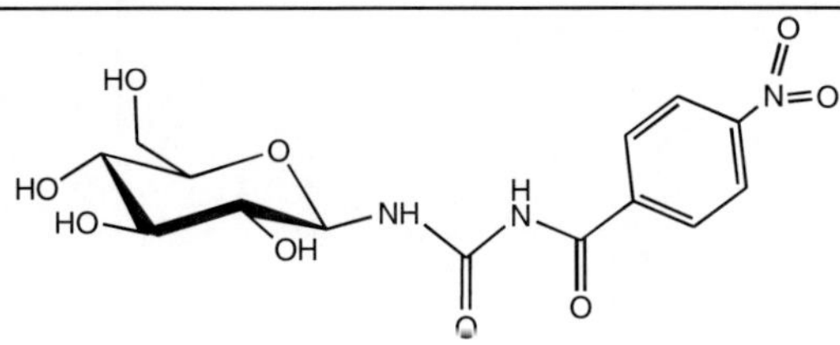	3.3 [27]
11	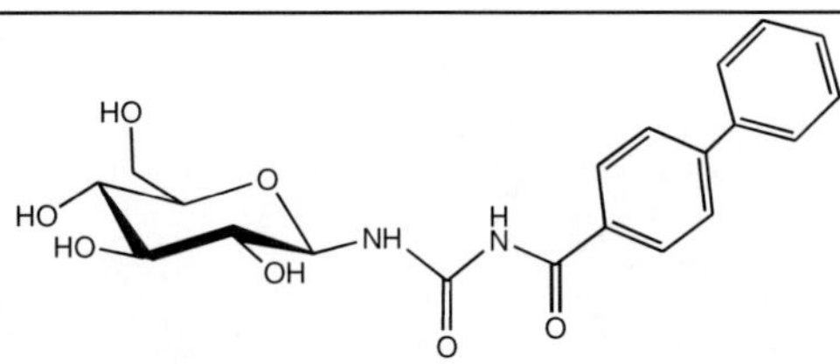	3.7 [27]
12	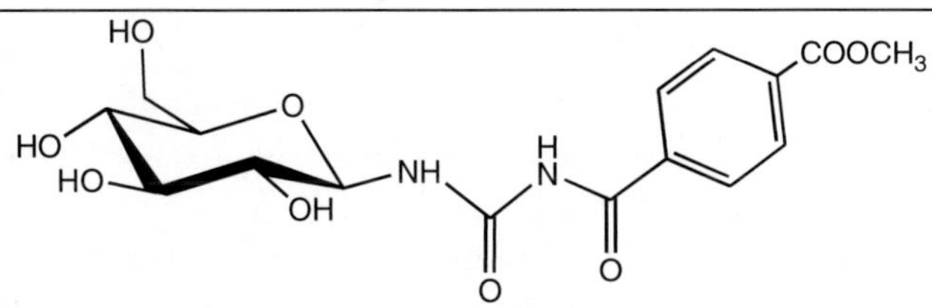	4.0 [27]

Table 1. (Continued)

	Chemical Structure	Ki (μM)
13	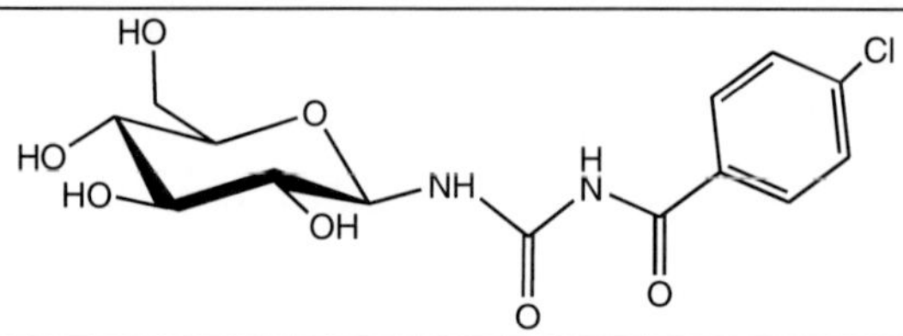	4.4 [27]
14	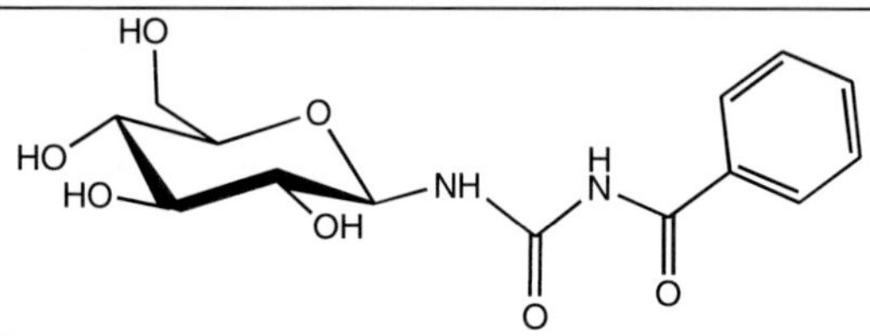	4.6 [27]
15	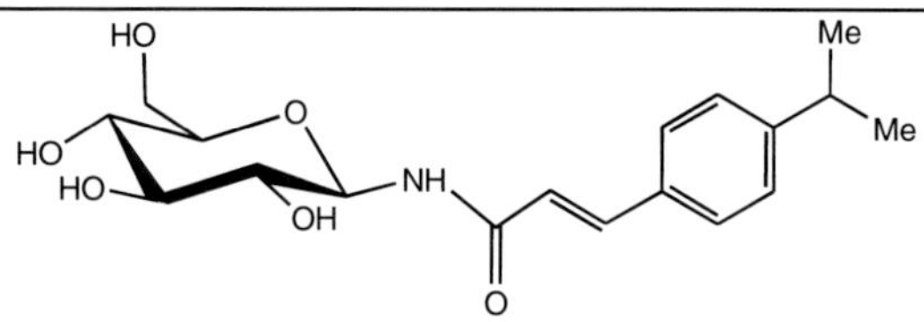	5.0
16	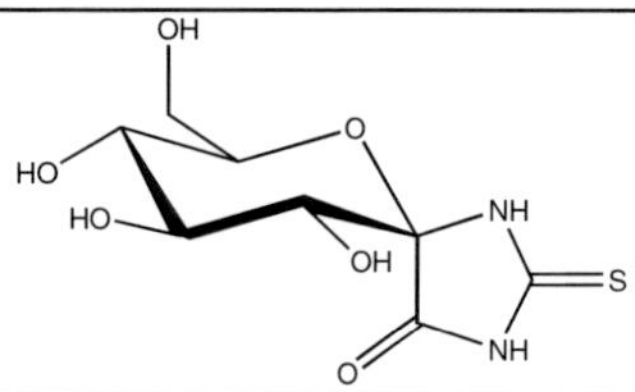	5.1 [27]
17	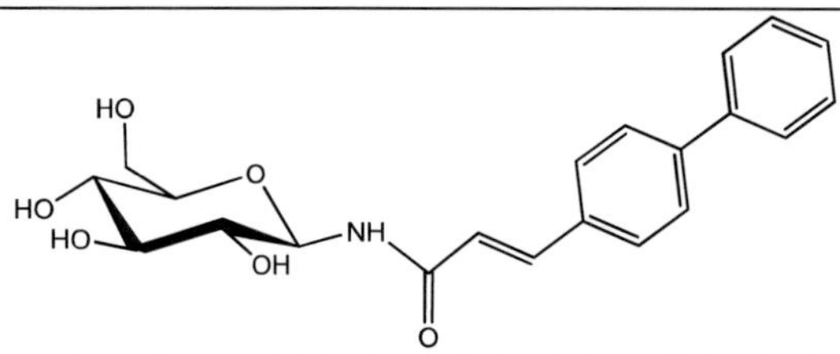	5.1

The GP catalytic site is a deep cavity located at the center of the molecule, 15 Å from the protein surface, and close to the essential cofactor PLP (Figure 1). This site has been probed with inhibitors (Table 1) that are trying to mimic the binding of the physiological inhibitor glucose, and they were designed on the basis of information derived from the crystal structure of the T-state GPb-α-D-glucose complex [22]. In previous studies [2] but also in a more recent one [23] any modification of the hydroxyl groups of the glucose molecule leads to a decrease in the potency with respect to glucose. This has been seen even in inhibitors that form additional hydrogen bond interactions with the natural cofactor, PLP. This paradox has been recently explained by Quantum Mechanics/Molecular Mechanics – Poisson Boltzmann Surface Area (QM/MM-PBSA) binding affinity calculations [23]. The QM/MM-PBSA calculations revealed that the desolvation energy cost for binding out-weigh the enthalpy gains from the extra contacts formed. Following the initial observations from the T-state GPb-α-D-glucose complex [22] attention has shifted towards molecules that have α- and β-substitutions at the anomeric *C*1 atom of the glucose moiety [14, 16] with the *β*-substitutions aiming to exploit the *β*-channel an empty space at the *β*-1-C configuration lined by both polar and non-polar groups. The most potent inhibitors for the catalytic site thus far have hydrophobic groups trying to fill the β-channel [13, 14, 24]. The common kinetic and structural features of these compounds are that they bind at the catalytic site and stabilise the T-state conformation (closed) of the 280s loop. The most efficient glucose derived inhibitors of GPb are presented in Table 1. Their efficiency has been attributed to the exploitation of extensive interactions with the *β*-channel.

A series of *C*5-alkynyl and alkylfurano[2,3-*d*]pyrimidine glucopyranonucleosides have been synthesized and studied for inhibition and binding to GPb. Most of these compounds had K_i values in the low μM range, with the most potent being 1-(*β*-D-glucopyranosyl)-5-ethynyluracil (Table 1). Structural studies revealed that inhibitors with a long *C*5-alkynyl group exploiting a large number of interactions within the β channel of the active site induce significant conformational changes of the 280s loop as compared to those with a shorter *C*5-alkynyl group which were the most potent [25]. This study highlighted the importance in the length of the aliphatic groups and the interactions within the β-channel showing the optimum length for potency. The best of these inhibitors had also a moderate effect on glycogenolysis in the cellular lever with an IC_{50} value of 291.4 μM [25]. Following that discovery, another study on the effect of the *C*-5 halogen substitution of glucopyranosyl nucleosides (1-(β-D-glucopyranosyl)5-X-uracil; X=Cl, Br, I) has led to the discovery of some of the most potent inhibitors of GP to date (Table 1). The structural basis of their inhibitory potential was the halogen's ability to form intermolecular interactions of electrostatic nature through the σ-hole phenomenon as revealed by X-ray crystallography and quantum mechanics [26]. These inhibitors did not show any cytotoxicity and they were moderately effective in *ex vivo* experiments. The main problem of their moderate activity at cellular level was attributed by adsorption, distribution, metabolism, excretion and toxicity (ADMET) property calculations, to their polar surface areas [26]. On another recent study [27], *N*-(4-Substituted-benzoyl)-*N*'-(β-D-glucopyranosyl) ureas (substituents: Me, Ph, Cl, OH, OMe, NO_2, NH_2,COOH, and COOMe) were synthesized and tested for inhibition of rabbit muscle glycogen phosphorylase b (RMGPb). Most of these compounds were low μM inhibitors and the most potent was the 4-methylbenzoyl compound (Table 1) with K_i = 2.3 μM. Crystallographic analyses showed that the analogues exploited the available space at the β-channel and induced an extended shift of the 280s loop. Molecular docking studies of selected GP inhibitors, has demonstrated the ability to calculate K_i values that correlate well with the experimentally measured values [15, 27]. Thus, a virtual screening of 1888 β-D-glucopyranose-NH-CO-R compounds where the diverse set of R groups was derived from acid chloride (RCOCl) molecules downloaded from the Zinc docking database [28] has been performed. Selected candidates from the virtual screening were then experimentally evaluated, with ADME property predictions allowing the selection of a set of *β*-D-glucopyranose-NH-CO-R ligands with drug-like properties as a criterion for further evaluation [29, 30]. Six selected candidates from the screening were then synthesized and their inhibitory potency was assessed both *in vitro* and *ex vivo*. This study produced inhibitors with K_i values in the range of low μM (Table 1) and provided the data for further structure based design of more potent inhibitors.

Recently, the GP inhibitor site has been also attracted some research interest especially using compounds from natural products. Thus, flavonoid derivatives [31-34], triterpenes [35], and benzoic acid derivatives [36, 37] have been found to display significant inhibitory potency. All these compounds bind at the inhibitor binding site blocking the entrance to the active site and have the potential for exploitation, in the making of biofunctional foods for type 2 diabetes patients.

PhK Overview

PhK is a large hexadecameric protein of MW 1.3 MDa comprising of four subunits α,β,γ, δ with stoichiometry $(\alpha\beta\gamma\delta)_4$. The α (MW=134.8KDa), β (MW=125.2 KDa) and δ (MW=17.7 KDa) are regulatory subunits integrating extracellular hormonal (through cAMP, Ca^{2+}) and neuronal stimuli (mediated through Ca^{2+}) controlling thus the activity of the catalytic γ (MW= 44.7 KDa). The catalytic γ subunit is comprised of an N-terminal catalytic region (kinase domain) and a C-terminal calmodulin binding region (Figure 2). The δ subunit is an intrinsic calmodulin molecule conferring calcium sensitivity to the enzyme. The α and β subunits are phosphorylated (by cAMP-dependent kinase) and controlled by extrinsic calmodulin and metabolites such as ADP [38-40]. PhK is found in all tissues with prevalence in muscle, brain and liver. PhK is associated with glycogen granules and along with other enzymes of glycogen metabolism, in brain mainly found in the astroglial cells, a cell lineage that plays a major role in brain homeostasis [41], and with myofibrils in muscle cells or associated with SR [42]. The different isoforms of the enzyme that exist in muscle and liver reflect the different destinations of Glc-1-P, first product of glycogenolysis. In muscle glycogen breakdown is coordinated with Ca^{2+} release (and hence muscle contraction) to provide a ready source of energy. In liver, glucose-1-phosphate is produced in response to low blood sugar levels. Hence, PhK can incorporate hormonal, neuronal and metabolic signals in a tissue-specific manner.

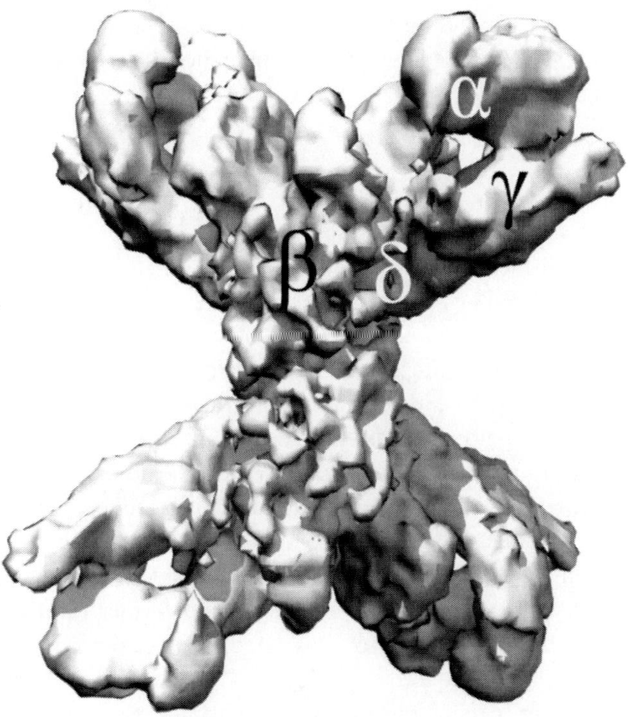

Figure 2. The cryo-electron microscopy 3D structure of PhK. The position of the different subunits are indicated in one of the monomers. Adapted from [53].

GP and glycogen synthase (GS) have been identified as the only in vivo substrates of PhK although various proteins have been reported that are phosphorylated by PhK *in vitro* including troponin I, troponin T, casein, histone H,1 neurogranin, neuromodulin its own β subunit and the Tau protein associated with Alzheimer's disease [43-49]. Analysis of phopshorylation sites of substrates revealed a consensus sequence of (R/K)XXS*(V/I)Z using the single letter amino acid code where S* is the serine phosphorylated and X and Z may beany amino acid, with increased activity if Z is an arginine [47, 50].

A catalytic mechanism of PhK has been proposed [51] according to which phosphorylation proceeds with rapid equilibrium binding of substrates (random kinetic mechanism) the chemical step is fast and the rate-limiting step is the dissociation of products and ADP release. The individual microscopic rate constants were estimated by viscosity kinetic measurements. The contributions to catalysis of conserved amino acids such the catalytic Asp149 have been probed by mutational studies in combination to crystallographic evidence [52]. The chemical step is consistent with the direct in line phopshoryl transfer supporting the notion of an associative mechanism as the transition state involves a pentavallent phosphorane where the apical bonds are weak and long.

Structural Studies of PhK

PhK attracted the scientific interest for decades· it was the first protein kinase to be discovered in 1955 by Fisher and Krebs, yet the structural basis of its regulation is poorly understood. Analytical structural studies on PhK were hindered by the enzyme's complexity. Nevertheless, single particle analysis by electron microscopy [53] allowed the reconstructions of 3D models of Phk holoenzyme from rabbit muscle at 9.9Å resolution. In initial [54, 55] negative stain micrographs three major forms were viewed: the dominant butterfly form consisting of two wings, the chalice form consisting of a cup and a stem and the tetrad form characterised by a rectangular shape structure with a stain - filled central cavity. 3D reconstructions by the random conical tilt method revealed a structure exhibiting 222 symmetry and further refinement of the butterfly form showed that it was consisted of two opposing lobes held together by two short oblique cross-bridges. The better resolved structure of 9.9 Å shows finer details in agreement with the previously mentioned structures. The overall dimensions were 270x225x160 Å, the lobe dimensions are 225x110 Å while the bridge length are 55 Å and width 45 Å. The angle between the lobes is 65° Two main differences in structural features have been reported in comparison of this model of active form (calcium activated PhK art pH 8.2) and previous cryo-model of inactive PhK [56]. The differences refer to the angle between the lobes which was reported 90 ° instead of 65 ° and the observation that the two bridges were resolved to four bridges with asymmetric connections to the two lobes were attributed to possible conformational changes during PhK activation. Several cross linking and immunoelectron microscopy studies contributed to the identification of the likely position of individual subunits. In the butterfly structure the tetramers (αβγδ) associate in a head to tail fashion with $(αβγδ)_2$ comprising each lobe. The α subunit is likely located at the exterior top of the lobes (Figure 2) as indicated by the location of epitopes for monoclonal antibodies raised against α subunits and further supported by cross linking studies [57, 58]. Its N-terminal interacts with β subunit [58] while its C-terminal interacts with catalytic γ subunit [58, 59]. The β subunit is located centrally at the lobes forming the connecting bridges as supported by various studies [40, 60-62] and found by cross linking to interact with α and γ subunits [58, 63]. Recently by CD studies the secondary structure of β subunit determined to be mostly helical while by using an multidisciplinary approach involving chemical cross linking, MS, threading and ab initio calculations an atomic model of the tertiary structure of β subunit was constructed (Nadeau et al., 2012). The model was docked to the central bridge region of PhK cryo- EM structure in support of the notion that β subunits act as a scaffold, positioning the lobes, upon which the other subunits are arrayed. The γ subunit is located at the interior of the lobes as indicated by antibody binding and further supported by molecular model fitting of the high resolution structure of its N-terminal (a.a. 1-299) catalytic domain PhKγ to the 9.9 Å structure [53]. Although there is no high resolution structural data on the C-terminal (a.a 300-378) of γ subunit the molecular fitting showed it is oriented towards the bridge. It is well established [64-66] that γ interacts with α subunit the C-terminal of γ subunit interacts with δ subunit which is found [67] to be located very close to the bridge. It is reported that calcium perturbs the interaction of α with the catalytic subunit suggesting that the flow of structural information from δ subunit to α is mediated by the C-terminal of the catalytic domain [68].

Although there are no detailed structural data for α and β subunits, various bioinformatics studies [69-71] predict that both subunits contain structurally similar domains A,B,C,D to other known enzymes. Indeed, in both subunits domains A and B are identified to be similar with the α/β barrel glucoamylase like GH15 proteins while the C and D domains are similar to calcineurin B proteins (CBL domains). CBL proteins are calcium binding proteins, members of the EF-hand family, regulating CIPKs or SOS2-like kinases [71]. The importance of these domains in PhK is demonstrated by the finding [71] that several mutations in these domains are related to X-linked liver glycogenosis. In a more recent study [72] for the β subunit, domains C and D are modeled as a single domain using not a CBL protein but a PP2AA template.

The 9.9 Å resolution structure allowed the reconstruction of a PhK/GPb model at 18 Å resolution [53]. Its overall dimensions were 310x250x200 Å greater than PhK dimensions (270x 225x 160 Å) due to the presence of GPb. The lobe dimensions showed a 25Å increase in length but no change in bridge length suggesting GPb binding towards the ends of the lobes. The glycogen site of GPb is exposed toward the top of PhK lobes. The volume of density for PhK –GPb indicates that only one subunit of GPb binds at each lobe and is positioned as to place the phosphorylatable Ser14 close to the position of γ subunit.

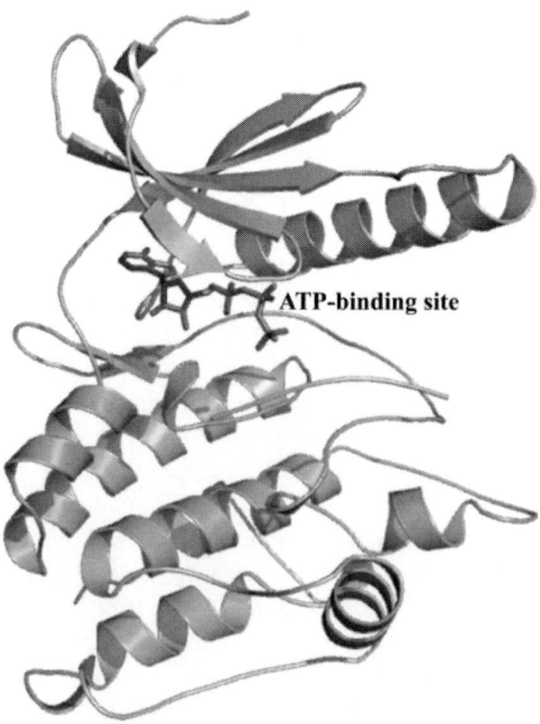

Figure 3. A schematic diagram of the γ-catalytic subunit of PhK with the ATP binding site position indicated by the ATP molecule. Adapted from [52].

Catalytic Subunit

The structures of a binary complex the kinase domain of γ catalytic subunit with ATP and the ternary complex with a non hydrolysable nucleotide analogue AMPPNP and an optimized peptide substrate and a mutant E182S of the catalytic subunit have been solved at a resolution of 2.6 2.3 and 2.4 Å respectively, by X-ray protein crystallography [51, 52, 73]. The kinase domain (residues 1-298) is constitutively active without the need of phosphorylation at the so called kinase activation segment (it has a Glu residue mimicking a phosphorylated Ser/Thr or Tyr) as other kinases, providing a good model for the study of catalysis and substrate recognition. Its structural analysis revealed (Figure 3) a common kinase fold consisting of two domains linked by a hinge region, an N-terminal domain (a.a 14-107) comprised mostly of antiparallel β sheets (strands β1-β9 and the helix aC) and a helical C terminal domain (a.a. 110-292). The ATP binding site is locates at the cleft between a glycine rich hairpin formed by strands β1 and β2 and the C-terminal domain. The C terminal domain also contains the activation segment of protein kinases between conserved sequences APE and DGF which in PhK coincides with the region between the amino acids 167-193. The catalytic Asp149 is buried surrounded by highly conserved Phe170, the peptide substrate, the activation segment residues and also Lys152 and Asn154 from the catalytic loop (a.a. 145-154) [51]. In the ternary complex, the optimized substrate peptide (MC Lys-Arg-Met-Met-Ser*-Phe-Phe-Leu-Phe) binds at the active site defined by an interdomain groove due to shape complementarity, hydropathy and potential. The interactions of peptide substrate with residues of the activation segment indicate the important role of correct conformation of the later for the substrate recognition. This notion is further supported by mutagenesis studies [51, 74] and the fact that mutations of the activation segment residues of the testis/liver isoform are related to autosomal glycogenosis [75].

PhK and ATP Competitive Inhibitors

PhK activates GP which catalyses the flux generating step in glycogen degradation [76] hence, it has a direct influence on blood glucose levels. In cell and animal models studies has been shown that hepatic insulin resistance is associated with impaired glycogen synthesis and increase in activity of GPa [77, 78].

Table 2. The chemical structures of the most potent inhibitors of the PhK ATP binding site discovered to date, together with the K_i values for each inhibitor [83]

	Chemical structure	K_i (nM)
Staurosporine	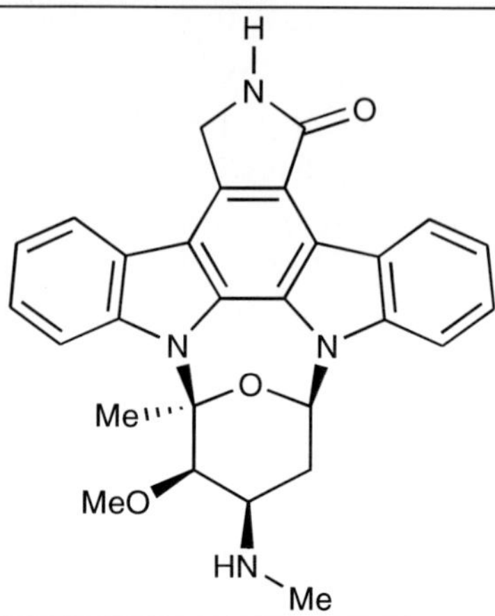	0.37
KT5720	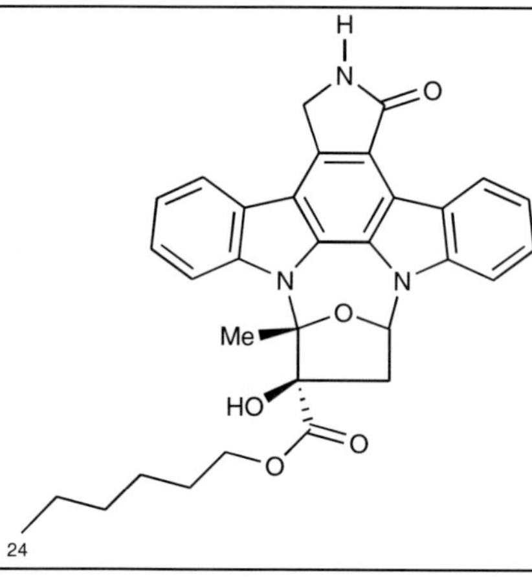	18.4
Indirubin 3-oxime	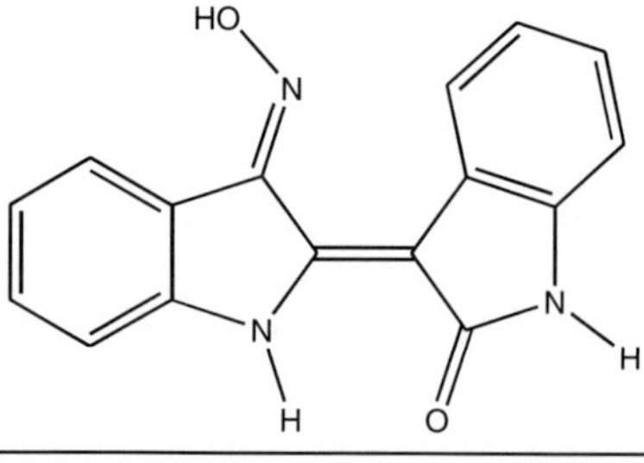	144

In addition, previous studies [79, 80] have shown that selective dephosphorylation of GPa by an indole carboxamide that favors the T (inactive) conformation of phosphorylase caused stimulation of glycogen synthesis. Inhibition of PhK by synthetic (modulators) inhibitors is expected to mimic metabolic conditions that decrease concentration of GPa leading thus to stimulation of glycogen synthesis. Despite the complexity of PhK structure the majority of reported PhK inhibitors bind at the ATP binding site. As protein kinases share a common architecture of their ATP binding site the challenge in the quest of ATP competitive inhibitors is not only potency but also selectivity. At least five pockets adjacent to the ATP binding site can be targeted by groups of chemical diversity in order to achieve inhibitor selectivity [81]. There have been several successful examples of potent and selective inhibitors in protein kinases, exploited in drug design (reviewed in [82]).

In a recent work [83], the binding of indirubin, staurosporine and their analogues indirubin 2-oxime and KT5720, respectively, in the ATP binding site was investigated by kinetics and *in silico* calculations involving docking, molecular dynamics and MM-GBSA approaches. Indirubin is a natural bis-indol antileukemic molecule from plants and mollusks and its derivatives are used in the treatment cancer, diabetes, and neurodegenerative diseases [84]. Staurosporine is a well known inhibitor of a number of protein kinases and it is too toxic for clinical use. However kinetic studies [85] revealed that its analogue KT5720 is a selective PhK inhibitor. Kinetic experiments using the constitutively active truncated PhKγt showed (Table 2) that indirubin-3-oxime, KT5720 and staurosporine have been identified as nM inhibitors of the PhK ATP binding site. The binding modes were predicted by *in silico* calculations and the potential for more selective PhK inhibition was established. The importance of bridging water molecules was revealed as the PhKγt-staurosporine complex had the greatest number of receptor-ligand hydrogen bonds and favorable electrostatic interactions while for complexes with the indirubin-3-oxime and the more selective KT5720 inhibitor there was a network of interchanging water molecules bridging inhibitor-enzyme contacts.

In the recent years new concepts were introduced in the exploitation of PhK as a molecular drug target. Bioinformatics studies predicted as aforementioned new functional domains in PhK such as the GH15-like domain in the regulatory α and β subunits. The validity of these predictions were tested experimentally when it was demonstrated that glycoamylase inhibitor acarbose binds strongly to PhK [K_D 3.7μM (pH 6.8) and 5.6 μM (pH 8.2)], causes conformational perturbations, and activates it [86]. Moreover, it has been shown [87] that there is a link between upregulated PhK gene expression and tumorigenesis connecting glycogen metabolism with tumor progression. Tumor progression is associated with pathological angiogenesis [88]. High throughput screening in zebrafish *in vivo* system and in human endothelial cell *in vitro* angiogenesis models have been used to identify novel antiangiogenic PhK inhibitors indicating a new role for PhK as a therapeutic target [87].

ACKNOWLEDGMENTS

VTS and DDL are grateful to the late Dr. Nikos G. Oikonomakos and Prof. Louise N. Johnson for introducing them to the GP and PhK projects. We also wish to thank our colleagues Joseph M. Hayes, Catherine Venien-Bryan, Dimitris Komiotis, Anna-Maria Psarra, Stella Manta, Spyros Zographos, Katerina Tsitsanou, Loranne Agius and George Archontis for their contributions to this project. This project was implemented under the "ARISTEIA" Action of the "Operational Programme Education and Lifelong Learning" and is co-funded by the European Social Fund (ESF) and National Resources. This work was also supported in part by the Postgraduate Programmes ‘‘Biotechnology-Quality assessment in Nutrition and the Environment”, ‘‘Application of Molecular Biology-Molecular Genetics-Molecular Markers”, Department of Biochemistry and Biotechnology, University of Thessaly.

REFERENCES

[1] Agius, L., New hepatic targets for glycaemic control in diabetes. *Best Pract. Res. Clin. Endocrinol. Metab.*, 2007 21, 587-605.

[2] Oikonomakos, N. G., Glycogen phosphorylase as a molecular target for type 2 diabetes therapy. *Curr. Protein. Pept. Sci.*, 2002 3, 561-586.

[3] Goldberg, J., Huang, H. B., Kwon, Y. G., Greengard, P., Nairn, A. C., and Kuriyan, J., Three-dimensional structure of the catalytic subunit of protein serine/threonine phosphatase-1. *Nature*, 1995 376, 745-753.

[4] Treadway, J. L., Mendys, P., and Hoover, D. J., Glycogen phosphorylase inhibitors for treatment of type 2 diabetes mellitus. *Expert Opin. Investig. Drugs*, 2001 10, 439-454.

[5] Carpino, P. A. and Goodwin, B., Diabetes area participation analysis: a review of companies and targets described in the 2008 - 2010 patent literature. *Expert Opin. Ther. Pat.*, 2010 20, 1627-1651.

[6] Newgard, C. B., Hwang, P. K., and Fletterick, R. J., The family of glycogen phosphorylases: structure and function. *Crit. Rev. Biochem. Mol. Biol.*, 1989 24, 69-99.

[7] Monod, J., Wyman, J., and Changeux, J. P., On the Nature of Allosteric Transitions: A Plausible Model. *J. Mol. Biol.*, 1965 12, 88-118.

[8] Barford, D. and Johnson, L. N., The allosteric transition of glycogen phosphorylase. *Nature*, 1989 340, 609-616.

[9] Barford, D. and Johnson, L. N., The molecular mechanism for the tetrameric association of glycogen phosphorylase promoted by protein phosphorylation. *Protein Sci.*, 1992 1, 472-493.

[10] Sprang, S. R., Acharya, K. R., Goldsmith, E. J., Stuart, D. I., Varvill, K., Fletterick, R. J., Madsen, N. B., and Johnson, L. N., Structural-Changes in Glycogen-Phosphorylase Induced by Phosphorylation. *Nature*, 1988 336, 215-221.

[11] Barford, D., Hu, S. H., and Johnson, L. N., Structural mechanism for glycogen phosphorylase control by phosphorylation and AMP. *J. Mol. Biol.*, 1991 218, 233-260.

[12] Sprang, S. R., Withers, S. G., Goldsmith, E. J., Fletterick, R. J., and Madsen, N. B., Structural Basis for the Activation of Glycogen Phosphorylase-B by Adenosine-Monophosphate. *Science*, 1991 254, 1367-1371.

[13] Oikonomakos, N. G. and Somsak, L., Advances in glycogen phosphorylase inhibitor design. *Curr. Opin. Investig. Drugs*, 2008 9, 379-395.

[14] Somsak, L., Czifrak, K., Toth, M., Bokor, E., Chrysina, E. D., Alexacou, K. M., Hayes, J. M., Tiraidis, C., Lazoura, E., Leonidas, D. D., Zographos, S. E., and Oikonomakos, N. G., New inhibitors of glycogen phosphorylase as potential antidiabetic agents. *Curr. Med. Chem.*, 2008 15, 2933-2983.

[15] Hayes, J. M. and Leonidas, D. D., Computation as a tool for glycogen phosphorylase inhibitor design. *Mini Rev. Med. Chem.*, 2010 10, 1156-1174.

[16] Somsak, L., Glucose derived inhibitors of glycogen phosphorylase. *Comptes Rendus Chimie*, 2011 14, 211-223.

[17] Agius, L., Physiological control of liver glycogen metabolism: lessons from novel glycogen phosphorylase inhibitors. *Mini Rev. Med. Chem.*, 2010 10, 1175-1187.

[18] Kurukulasuriya, R., Link, J. T., Madar, D. J., Pei, Z., Richards, S. J., Rohde, J. J., Souers, A. J., and Szczepankiewicz, B. G., Potential drug targets and progress towards pharmacologic inhibition of hepatic glucose production. *Curr. Med. Chem.*, 2003 10, 123-153.

[19] Kurukulasuriya, R., Link, J. T., Madar, D. J., Pei, Z., Rohde, J. J., Richards, S. J., Souers, A. J., and Szczepankiewicz, B. G., Prospects for pharmacologic inhibition of hepatic glucose production. *Curr. Med. Chem.*, 2003 10, 99-121.

[20] Board, M., Hadwen, M., and Johnson, L. N., Effects of novel analogues of D-glucose on glycogen phosphorylase activities in crude extracts of liver and skeletal muscle. *Eur. J. Biochem.*, 1995 228, 753-761.

[21] Rath, V. L., Ammirati, M., LeMotte, P. K., Fennell, K. F., Mansour, M. N., Danley, D. E., Hynes, T. R., Schulte, G. K., Wasilko, D. J., and Pandit, J., Activation of human liver glycogen phosphorylase by alteration of the secondary structure and packing of the catalytic core. *Mol. Cell*, 2000 6, 139-148.

[22] Martin, J. L., Veluraja, K., Ross, K., Johnson, L. N., Fleet, G. W. J., Ramsden, N. G., Bruce, I., Orchard, M. G., Oikonomakos, N. G., Papageorgiou, A. C., Leonidas, D. D., and Tsitoura, H. S., Glucose analogue inhibitors of glycogen phosphorylase: The design of potential drugs for diabetes. *Biochemistry (USA)*, 1991 30, 10101-10116.

[23] Manta, S., Xipnitou, A., Kiritsis, C., Kantsadi, A. L., Hayes, J. M., Skamnaki, V. T., Lamprakis, C., Kontou, M., Zoumpoulakis, P., Zographos, S. E., Leonidas, D. D., and Komiotis, D., 3'-Axial CH(2) OH Substitution on Glucopyranose does not Increase Glycogen Phosphorylase Inhibitory Potency. QM/MM-PBSA Calculations Suggest Why. *Chem. Biol. Drug Des.*, 2012 79, 663-673.

[24] Papageorgiou, A. C., Oikonomakos, N. G., Leonidas, D. D., Bernet, B., Beer, D., and Vasella, A., The binding of D-gluconohydroximo-1,5-lactone to glycogen phosphorylase. Kinetic, ultracentrifugation and crystallographic studies. *Biochem. J.*, 1991 274, 329-338.

[25] Kantsadi, A. L., Manta, S., Psarra, A. M., Dimopoulou, A., Kiritsis, C., Parmenopoulou, V., Skamnaki, V. T., Zoumpoulakis, P., Zographos, S. E., Leonidas, D. D., and Komiotis, D., The binding of C5-alkynyl and alkylfurano[2,3-d]pyrimidine glucopyranonucleosides to glycogen

phosphorylase b: synthesis, biochemical and biological assessment. *Eur. J. Med. Chem.*, 2012 54, 740-749.

[26] Kantsadi, A. L., Hayes, J. M., Manta, S., Skamnaki, V. T., Kiritsis, C., Psarra, A. M., Koutsogiannis, Z., Dimopoulou, A., Theofanous, S., Nikoleousakos, N., Zoumpoulakis, P., Kontou, M., Papadopoulos, G., Zographos, S. E., Komiotis, D., and Leonidas, D. D., The sigma-Hole Phenomenon of Halogen Atoms Forms the Structural Basis of the Strong Inhibitory Potency of C5 Halogen Substituted Glucopyranosyl Nucleosides towards Glycogen Phosphorylase b. *ChemMedChem*, 2012 7, 722-732.

[27] Nagy, V., Felfoldi, N., Konya, B., Praly, J. P., Docsa, T., Gergely, P., Chrysina, E. D., Tiraidis, C., Kosmopoulou, M. N., Alexacou, K. M., Konstantakaki, M., Leonidas, D. D., Zographos, S. E., Oikonomakos, N. G., Kozmon, S., Tvaroska, I., and Somsak, L., N-(4-Substituted-benzoyl)-N'-(beta-d-glucopyranosyl)ureas as inhibitors of glycogen phosphorylase: Synthesis and evaluation by kinetic, crystallographic, and molecular modelling methods. *Bioorg. Med. Chem.*, 2012 20, 1801-1816.

[28] Irwin, J. J. and Shoichet, B. K., ZINC--a free database of commercially available compounds for virtual screening. *J. Chem. Inf. Model*, 2005 45, 177-182.

[29] Van de Waterbeemd, H., From in vivo to in vitro/in silico ADME: progress and challenges. *Expert Opin. Drug Metab. Toxicol.*, 2005 1, 1-4.

[30] Thompson, T. N., Early ADME in support of drug discovery: the role of metabolic stability studies. *Curr. Drug Metab.*, 2000 1, 215-241.

[31] Kaiser, A., Nishi, K., Gorin, F. A., Walsh, D. A., Bradbury, E. M., and Schnier, J. B., The cyclin-dependent kinase (CDK) inhibitor flavopiridol inhibits glycogen phosphorylase. *Arch. Biochem. Biophys.*, 2001 386, 179-187.

[32] Kato, A., Kamiyama, O., Sanae, F., Ikeda, K., Higashi, Y., Minami, Y., Asano, N., and Adachi, I., In vitro inhibition of alpha-glucosidases and glycogen phosphorylase by catechin gallates in green tea. *Food Chem.*, 2010 122, 1061-1066.

[33] Oikonomakos, N. G., Schnier, J. B., Zographos, S. E., Skamnaki, V. T., Tsitsanou, K. E., and Johnson, L. N., Flavopiridol inhibits glycogen phosphorylase by binding at the inhibitor site. *J. Biol. Chem.*, 2000 275, 34566-34573.

[34] Tsitsanou, K. E., Hayes, J. M., Keramioti, M., Mamais, M., Oikonomakos, N. G., Kato, A., Leonidas, D. D., and Zographos, S. E., Sourcing the affinity of flavonoids for the glycogen phosphorylase inhibitor site via crystallography, kinetics and QM/MM-PBSA binding studies: Comparison of chrysin and flavopiridol. *Food Chem. Toxicol.*, 2012, in press.

[35] Wen, X., Sun, H., Liu, J., Cheng, K., Zhang, P., Zhang, L., Hao, J., Ni, P., Zographos, S. E., Leonidas, D. D., Alexacou, K. M., Gimisis, T., Hayes, J. M., and Oikonomakos, N. G., Naturally occurring pentacyclic triterpenes as inhibitors of glycogen phosphorylase: synthesis, structure-activity relationships, and X-ray crystallographic studies. *J. Med. Chem.*, 2008 51, 3540-3554.

[36] Li, J., Lu, Y., Su, X., Li, F., She, Z., He, X., and Lin, Y., A norsesquiterpene lactone and a benzoic acid derivative from the leaves of Cyclocarya paliurus and their glucosidase and glycogen phosphorylase inhibiting activities. *Planta Med.*, 2008 74, 287-289.

[37] Li, S., Li, J., Guan, X. L., Deng, S. P., Li, L. Q., Tang, M. T., Huang, J. G., Chen, Z. Z., and Yang, R. Y., Hypoglycemic effects and constituents of the barks of Cyclocarya paliurus and their inhibiting activities to glucosidase and glycogen phosphorylase. *Fitoterapia*, 2011 82, 1081-1085.

[38] Brushia, R. J. and Walsh, D. A., Phosphorylase kinase: the complexity of its regulation is reflected in the complexity of its structure. *Front. Biosci.*, 1999 4, D618-641.

[39] Heilmeyer, L. M., Jr., Molecular basis of signal integration in phosphorylase kinase. *Biochim. Biophys. Acta*, 1991 1094, 168-174.

[40] Pickett-Gies, C. A. and Walsh, D. A., *Glycogen phosphorylase kinase.*, in *The Enzymes*, P. D. Boyer and E. G. Krebs, Editors. 1986, Academic Press. p. 396-459.

[41] Psarra, A. M., Pfeiffer, B., Giannakopoulou, M., Sotiroudis, T. G., Stylianopoulou, F., and Hamprecht, B., Immunocytochemical localization of glycogen phosphorylase kinase in rat brain sections and in glial and neuronal primary cultures. *J. Neurocytol.*, 1998 27, 779-790.

[42] Thieleczek, R., Behle, G., Messer, A., Varsanyi, M., Heilmeyer, L. M., Jr., and Drenckhahn, D., Localization of phosphorylase kinase subunits at the sarcoplasmic reticulum of rabbit skeletal muscle by monoclonal and polyclonal antibodies. *Eur. J. Cell Biol.*, 1987 44, 333-340.

[43] Akatsuka, A., Singh, T. J., and Huang, K. P., Phosphorylation of rat liver glycogen synthase by phosphorylase kinase. *J. Biol. Chem.*, 1984 259, 7878-7883.

[44] Kilimann, M. W. and Heilmeyer, L. M., Jr., Multiple activities on phosphorylase kinase. 1. Characterization of three partial activities by their response to calcium ion, magnesium ion, pH, and ammonium chloride and effect of activation by phosphorylation and proteolysis. *Biochemistry*, 1982 21, 1727-1734.

[45] DePaoli-Roach, A. A., Bingham, E. W., and Roach, P. J., Phosphorylase kinase from rabbit skeletal muscle: phosphorylation of kappa-casein. *Arch. Biochem. Biophys.*, 1981 212, 229-236.

[46] Tabuchi, H., Hashimoto, E., Nakamura, S., Yamamura, H., and Nishizuka, Y., Phosphorylation of calf thymus H1 histone by muscle glycogen phosphorylase kinase. *J. Biochem.*, 1981 89, 1433-1437.

[47] Pearson, R. B. and Kemp, B. E., Protein kinase phosphorylation site sequences and consensus specificity motifs: tabulations. *Methods Enzymol.*, 1991 200, 62-81.

[48] Paudel, H. K., The regulatory Ser262 of microtubule-associated protein tau is phosphorylated by phosphorylase kinase. *J. Biol. Chem.*, 1997 272, 1777-1785.

[49] Paudel, H. K., Zwiers, H., and Wang, J. H., Phosphorylase kinase phosphorylates the calmodulin-binding regulatory regions of neuronal tissue-specific proteins B-50 (GAP-43) and neurogranin. *J. Biol. Chem.*, 1993 268, 6207-6213.

[50] Graves, D. J., Use of peptide substrates to study the specificity of phosphorylase kinase phosphorylation. *Methods Enzymol.*, 1983 99, 268-278.

[51] Skamnaki, V. T., Owen, D. J., Noble, M. E., Lowe, E. D., Lowe, G., Oikonomakos, N. G., and Johnson, L. N., Catalytic mechanism of phosphorylase kinase probed by mutational studies. *Biochemistry*, 1999 38, 14718-14730.

[52] Lowe, E. D., Noble, M. E., Skamnaki, V. T., Oikonomakos, N. G., Owen, D. J., and Johnson, L. N., The crystal structure of a phosphorylase kinase peptide substrate complex: kinase substrate recognition. *EMBO J.*, 1997 16, 6646-6658.

[53] Venien-Bryan, C., Jonic, S., Skamnaki, V., Brown, N., Bischler, N., Oikonomakos, N. G., Boisset, N., and Johnson, L. N., The structure of phosphorylase kinase holoenzyme at 9.9 angstroms resolution and

location of the catalytic subunit and the substrate glycogen phosphorylase. *Structure*, 2009 17, 117-127.

[54] Venien-Bryan, C., Lowe, E. M., Boisset, N., Traxler, K. W., Johnson, L. N., and Carlson, G. M., Three-dimensional structure of phosphorylase kinase at 22 A resolution and its complex with glycogen phosphorylase b. *Structure*, 2002 10, 33-41.

[55] Nadeau, O. W., Carlson, G. M., and Gogol, E. P., A Ca(2+)-dependent global conformational change in the 3D structure of phosphorylase kinase obtained from electron microscopy. *Structure*, 2002 10, 23-32.

[56] Nadeau, O. W., Gogol, E. P., and Carlson, G. M., Cryoelectron microscopy reveals new features in the three-dimensional structure of phosphorylase kinase. *Protein Sci.*, 2005 14, 914-920.

[57] Wilkinson, D. A., Marion, T. N., Tillman, D. M., Norcum, M. T., Hainfeld, J. F., Seyer, J. M., and Carlson, G. M., An epitope proximal to the carboxyl terminus of the alpha-subunit is located near the lobe tips of the phosphorylase kinase hexadecamer. *J. Mol. Biol.*, 1994 235, 974-982.

[58] Nadeau, O. W. and Carlson, G. M., Zero length conformation-dependent cross-linking of phosphorylase kinase subunits by transglutaminase. *J. Biol. Chem.*, 1994 269, 29670-29676.

[59] Harris, W. R., Malencik, D. A., Johnson, C. M., Carr, S. A., Roberts, G. D., Byles, C. A., Anderson, S. R., Heilmeyer, L. M., Jr., Fischer, E. H., and Crabb, J. W., Purification and characterization of catalytic fragments of phosphorylase kinase gamma subunit missing a calmodulin-binding domain. *J. Biol. Chem.*, 1990 265, 11740-11745.

[60] Wilkinson, D. A., Norcum, M. T., Fizgerald, T. J., Marion, T. N., Tillman, D. M., and Carlson, G. M., Proximal regions of the catalytic gamma and regulatory beta subunits on the interior lobe face of phosphorylase kinase are structurally coupled to each other and with enzyme activation. *J. Mol. Biol.*, 1997 265, 319-329.

[61] Trempe, M. R., Carlson, G. M., Hainfeld, J. F., Furcinitti, P. S., and Wall, J. S., Analyses of phosphorylase kinase by transmission and scanning transmission electron microscopy. *J. Biol. Chem.*, 1986 261, 2882-2889.

[62] Ayers, N. A., Nadeau, O. W., Read, M. W., Ray, P., and Carlson, G. M., Effector-sensitive cross-linking of phosphorylase b kinase by the novel cross-linker 4-phenyl-1,2,4-triazoline-3,5-dione. *Biochem. J.*, 1998 331 (Pt 1), 137-141.

[63] Nadeau, O. W., Anderson, D. W., Yang, Q., Artigues, A., Paschall, J. E., Wyckoff, G. J., McClintock, J. L., and Carlson, G. M., Evidence for the location of the allosteric activation switch in the multisubunit phosphorylase kinase complex from mass spectrometric identification of chemically crosslinked peptides. *J. Mol. Biol.*, 2007 365, 1429-1445.

[64] Dasgupta, M. and Blumenthal, D. K., Characterization of the regulatory domain of the gamma-subunit of phosphorylase kinase. The two noncontiguous calmodulin-binding subdomains are also autoinhibitory. *J. Biol. Chem.*, 1995 270, 22283-22289.

[65] Cook, A. G., Johnson, L. N., and McDonnell, J. M., Structural characterization of Ca2+/CaM in complex with the phosphorylase kinase PhK5 peptide. *FEBS J.*, 2005 272, 1511-1522.

[66] Dasgupta, M., Honeycutt, T., and Blumenthal, D. K., The gamma-subunit of skeletal muscle phosphorylase kinase contains two noncontiguous domains that act in concert to bind calmodulin. *J. Biol. Chem.*, 1989 264, 17156-17163.

[67] Traxler, K. W., Norcum, M. T., Hainfeld, J. F., and Carlson, G. M., Direct visualization of the calmodulin subunit of phosphorylase kinase via electron microscopy following subunit exchange. *J. Struct. Biol.*, 2001 135, 231-238.

[68] Rice, N. A., Nadeau, O. W., Yang, Q., and Carlson, G. M., The calmodulin-binding domain of the catalytic gamma subunit of phosphorylase kinase interacts with its inhibitory alpha subunit: evidence for a Ca2+ sensitive network of quaternary interactions. *J. Biol. Chem.*, 2002 277, 14681-14687.

[69] Pallen, M. J., Glucoamylase like domains in the alpha- and beta-subunits of phosphorylase kinase. *Protein Sci.*, 2003 12, 1804-1807.

[70] Carriere, C., Jonic, S., Mornon, J. P., and Callebaut, I., 3D mapping of glycogenosis-causing mutations in the large regulatory alpha subunit of phosphorylase kinase. *Biochim. Biophys. Acta*, 2008 1782, 664-670.

[71] Carriere, C., Mornon, J. P., Venien-Bryan, C., Boisset, N., and Callebaut, I., Calcineurin B-like domains in the large regulatory alpha/beta subunits of phosphorylase kinase. *Proteins*, 2008 71, 1597-1606.

[72] Nadeau, O. W., Lane, L. A., Xu, D., Sage, J., Priddy, T. S., Artigues, A., Villar, M. T., Yang, Q., Robinson, C. V., Zhang, Y., and Carlson, G. M., Structure and location of the regulatory beta subunits in the (alphabetagammadelta)4 phosphorylase kinase complex. *J. Biol. Chem.*, 287, 36651-36661.

[73] Owen, D. J., Noble, M. E., Garman, E. F., Papageorgiou, A. C., and Johnson, L. N., Two structures of the catalytic domain of phosphorylase kinase: an active protein kinase complexed with substrate analogue and product. *Structure*, 1995 3, 467-482.

[74] Skamnaki, V. T. and Oikonomakos, N. G., Kinetic characterization of the double mutant R148A/E182S of glycogen phosphorylase kinase catalytic subunit: the role of the activation loop. *J. Protein Chem.*, 2000 19, 499-505.

[75] Maichele, A. J., Burwinkel, B., Maire, I., Sovik, O., and Kilimann, M. W., Mutations in the testis/liver isoform of the phosphorylase kinase gamma subunit (PHKG2) cause autosomal liver glycogenosis in the gsd rat and in humans. *Nat. Genet.*, 1996 14, 337-340.

[76] Bollen, M., Keppens, S., and Stalmans, W., Specific features of glycogen metabolism in the liver. *Biochem. J.*, 1998 336 (Pt 1), 19-31.

[77] Arden, C., Green, A. R., Hampson, L. J., Aiston, S., Harndahl, L., Greenberg, C. C., Brady, M. J., Freeman, S., Poucher, S. M., and Agius, L., Increased sensitivity of glycogen synthesis to phosphorylase-a and impaired expression of the glycogen-targeting protein R6 in hepatocytes from insulin-resistant Zucker fa/fa rats. *FEBS J.*, 2006 273, 1989-1999.

[78] Aiston, S., Peak, M., and Agius, L., Impaired glycogen synthesis in hepatocytes from Zucker fatty fa/fa rats: the role of increased phosphorylase activity. *Diabetologia*, 2000 43, 589-597.

[79] Aiston, S., Coghlan, M. P., and Agius, L., Inactivation of phosphorylase is a major component of the mechanism by which insulin stimulates hepatic glycogen synthesis. *Eur. J. Biochem.*, 2003 270, 2773-2781.

[80] Hampson, L. J. and Agius, L., Increased potency and efficacy of combined phosphorylase inactivation and glucokinase activation in control of hepatocyte glycogen metabolism. *Diabetes*, 2005 54, 617-623.

[81] Liao, J. J., Molecular recognition of protein kinase binding pockets for design of potent and selective kinase inhibitors. *J. Med. Chem.*, 2007 50, 409-424.

[82] Johnson, L. N., Protein kinase inhibitors: contributions from structure to clinical compounds. *Q. Rev. Biophys.*, 2009 42, 1-40.

[83] Hayes, J. M., Skamnaki, V. T., Archontis, G., Lamprakis, C., Sarrou, J., Bischler, N., Skaltsounis, A. L., Zographos, S. E., and Oikonomakos, N. G., Kinetics, in silico docking, molecular dynamics, and MM-GBSA binding studies on prototype indirubins, KT5720, and staurosporine as phosphorylase kinase ATP-binding site inhibitors: the role of water molecules examined. *Proteins*, 2011 79, 703-719.

[84] Jautelat, R., Brumby, T., Schafer, M., Briem, H., Eisenbrand, G., Schwahn, S., Kruger, M., Lucking, U., Prien, O., and Siemeister, G., From the insoluble dye indirubin towards highly active, soluble CDK2-inhibitors. *Chembiochem*, 2005 6, 531-540.

[85] Davies, S. P., Reddy, H., Caivano, M., and Cohen, P., Specificity and mechanism of action of some commonly used protein kinase inhibitors. *Biochem. J.*, 2000 351, 95-105.

[86] Nadeau, O. W., Liu, W., Boulatnikov, I. G., Sage, J. M., Peters, J. L., and Carlson, G. M., The glucoamylase inhibitor acarbose is a direct activator of phosphorylase kinase. *Biochemistry*, 2010 49, 6505-6507.

[87] Camus, S., Quevedo, C., Menendez, S., Paramonov, I., Stouten, P. F., Janssen, R. A., Rueb, S., He, S., Snaar-Jagalska, B. E., Laricchia-Robbio, L., and Izpisua Belmonte, J. C., Identification of phosphorylase kinase as a novel therapeutic target through high-throughput screening for anti-angiogenesis compounds in zebrafish. *Oncogene*, 2012 31, 4333-4342.

[88] Carmeliet, P. and Jain, R. K., Angiogenesis in cancer and other diseases. *Nature*, 2000 407, 249-257.

[89] Benltifa, M., Hayes, J. M., Vidal, S., Gueyrard, D., Goekjian, P. G., Praly, J.-P., Kizilis, G., Tiraidis, C., Alexacou, K. M., Chrysina, E. D., Zographos, S. E., Leonidas, D. D., Archontis, G., and Oikonomakos, N. G., Glucose-based spiro-isoxazolines: a new family of potent glycogen phosphorylase inhibitors. *Bioorg. Med. Chem.*, 2009 17, 7368-7380.

INDEX

A

H

I

N

O

P

Q

R

S

W

Y